100 + Fun Ideas

Science Investigations

in the Primary Classroom

Anita Loughrey

Other Titles in the 100+ Fun Ideas series:

Practising Modern Foreign Languages in the Primary Classroom	978-1-903853-98-6
Art Activities	978-1-905780-33-4
Playground Games	978-1-905780-40-2
Transition Times	978-1-905780-34-1
Wet Playtimes	978-1-905780-32-7

Published by Brilliant Publications
Unit 10, Sparrow Hall Farm
Edlesborough, Dunstable
Bedfordshire, LU6 2ES, UK
www.brilliantpublications.co.uk

Sales and stock enquiries:
Tel: 01202 712910
Fax: 0845 1309300
E-mail: brilliant@bebc.co.uk
General information enquiries:
Tel: 01525 222292

The name Brilliant Publications and the logo are registered trademarks.

Written by Anita Loughrey
Illustrated by Pat Murray

ISBN 978-1-905780-35-8
First printed and published in the UK in 2009
10 9 8 7 6 5 4 3 2 1

Contents

Preface

How can you teach the basics of sound scientific investigation, while keeping your endeavours fun and relevant? That dilemma has motivated me to create this book, which is full of a huge range of exciting ideas for science classroom experiments, all of which have been tried and tested over my 17 years as a primary school teacher.

With the help of this book, children are encouraged to:

- Pose questions
- Make predictions
- Decide what evidence to collect
- Design a fair test
- Observe
- Record these observations appropriately and accurately
- Indicate whether a prediction was valid
- Explain their findings in scientific terms

The book comprises three main units, with sub-sections as follows:

- Life processes and living things – Humans and other animals
  – Living things in their environment
- Materials and their properties – Grouping and classifying
- Physical processes – Light and its properties
  – Sound and its properties
  – Forces and motion
  – Electricity

The activities require a minimum of preparation and use only the simplest of science equipment. As with all practical activities, safety is of utmost importance, and safety recommendations have been added where relevant. The variety of activities support the opportunity to develop your children's skills of scientific enquiry. So use them, adapt them, make them your own – but above all, enjoy them.

Anita Loughrey

Introduction

Group work
The majority of the investigations in this book suggest that children should work with a partner or in small groups. This will allow the children to share their experiences and to consider different points of view and solutions. I strongly believe group co-operation helps to develop children's social skills, and in particular those of shared leadership, communication, trust and conflict management. This is an important step towards the children's development into mature and responsible citizens.

Try to avoid social stereotypes such as boys manipulating materials and girls recording results. By careful observation of the groups' interactions, you can ensure individual children participate in all aspects of the investigations. I suggest you keep a record of the groups the children have worked in, so that by the end of the academic year each child has worked with as many others as possible.

Questioning
The questions you ask are very important for reinforcing the scientific concepts you are trying to teach and in helping each child to think carefully and clarify their own ideas. Allow time for them to think before they answer, and listen carefully to their responses so that you can build on their understanding by asking more questions to highlight scientific facts.

Science is all about asking questions about what is happening and finding the answers in a safe and logical way. This is why each investigation has been devised with a specific question for the children to answer.

Observation
Curiosity is the key to science. It is the desire to know why things happen and how both living and non-living things work. Therefore, science involves careful observation of what is going on around us.

Before each science investigation is conducted, each child should make a prediction about what they think will happen. This is good practice.

Emphasize how important it is to make careful observations and to keep accurate records of what happens during their investigations. The easiest way for them to record what they have done is to write their observations straight onto a chart. Such charts are useful in helping the children make their findings clear in an easy and immediately understandable way. This is why examples of charts, which could be used and adapted by the children, have been included in the book.

I have found young children often mix up results and conclusion. It is necessary to explain that the results are simply a record of their observations, which can be displayed by charts and graphs. The conclusion is what these observations tell us about the investigation. It is a good idea to start the conclusion with whether or not their prediction was correct. Talk about/discuss and encourage them to decide if their investigations were fair and how they could be improved.

Planning

The secret of carrying out meaningful investigations is to ensure careful planning.

Opportunities have been provided in this resource book for the children to devise and plan their own investigations. It is important to emphasize that, when they are planning an experiment, only one variable should change at a time; everything else must be left the same to ensure it is a fair test, so their conclusions will be valid.

Ourselves

1. Can you recognize objects from touch?

You will need: box or bag; blindfold; selection of objects young children would recognize (eg wet sponge, toy car, teddy, plastic brick, crayons, ice, stone etc); piece of fabric large enough to cover objects not in the bag or box.

Allow each group to secretly choose a selection of three or four objects to be placed in the box or bag at random.

Each group should challenge another group to recognize the chosen objects.

Encourage the groups to take it in turns to place an object into the bag or box. Remind the children to keep all objects hidden.

Blindfold each person in the group in turn and let them feel the object.

Allow them to have three guesses to find out what their allocated object is before they are shown it.

Which group recognized the most objects correctly from touch alone?

Talk about/discuss:

Ask the children what clues they used to recognize each object.

Explain that there are tiny nerves in our skin that tell us if surfaces are rough or smooth, hard or soft, wet or dry and hot or cold.

2. Can you recognize things from smell?

You will need: pots with lids (eg film cases); selection of items that smell strong and that young children would recognize (eg vinegar, onion, lemon, chocolate); several sheets of A1 paper; marker pens.

- Prepare a selection of different-smelling pots before the lesson begins. Label the pots 1, 2, 3 etc and have enough pots for every group to have one set each, with the same item in each corresponding pot.
- Suggest they take it in turns to sniff each pot and write or draw what they think they can smell.
- Encourage each group to record their guesses for each pot on large sheets of paper.
- Allow time for them to discuss each pot and come to a group consensus.
- At the end of the session, tell the children what was in each pot and see how many got each item correct.

Talk about/discuss:

- Ask the children if each pot smelt a lot different.
- Did they recognize all the smells?
- Explain that the sense of smell is strongest at birth. This is so that babies can recognize their mother before their eyesight has fully developed.

Safety:

✦ Do not forget to ask parents/carers if the children have any allergies before allowing the children to sniff/touch the pots.

3. Is the tallest person the oldest?

You will need: tape measure; metre ruler; chalk; large pieces of paper or playground.

In pairs, ask the children to draw around themselves on large sheets of paper. (If it's a nice day, they could draw around themselves outside on the playground using chalk.)

Encourage them to measure each other's outlines using non-standard measures such as lengths in feet, hands, spans, books, etc.

Some children may be ready to use standard measures with tape measure and rulers.

Compare their measured heights.

Discuss who are the tallest children and line them up in order of their birthdays.

Remember, the oldest children in the class will be those with their birthdays in September. Also be aware that some children may be sensitive about their height.

Talk about/discuss:

Have a class discussion about the subject.

Were the oldest children in the class the tallest?

Was this what was expected? Compare their findings with their initial hypothesis.

More ideas:

The data collected could be used to construct a simple ICT database.

4. Are everybody's hands the same?

You will need: ruler; coloured crayons; paint/ink.

- ✦ In small groups, the children should examine each other's hands. Are the lines on the palms of their hands the same?
- ✦ Encourage the children to use rulers to measure their handspan from the tip of their thumb to the tip of their little finger.
- ✦ Draw around each other's hands and cut the drawings out.
- ✦ Compare the sizes of their cut-out hands. These could be decorated and put on display.
- ✦ Use paint, or washable ink if available, to make fingerprints with both hands and compare their prints with those of other people in the group.
- ✦ Display labelled illustrations of different types of fingerprints, eg 'whorl', 'loop', 'arch' etc.

Talk about/discuss:

- ✦ Who has the largest hands?
- ✦ Who has the smallest hands?
- ✦ Were their fingerprints the same?
- ✦ Explain that everyone is unique and different. Tell the children this is why fingerprints can be used to identify people.

Note:
There may be a sensitive issue here as some parents may have reservations about their children's fingerprints being taken. Make sure this is not a problem before engaging in this activity and if necessary reassure parents that no permanent records of fingerprints will be kept.

5. Does the tallest person have the biggest feet?

You will need: tape measures and rulers; chalk; a copy of the chart for each pair; shoe size information.

As a homework task, ask the children to find out their shoe size.

Fill in the chart.

Name	Shoe size	Height

In pairs, the children should measure how tall they are by standing against a wall with their feet flat on the floor and marking their height with chalk. (Be aware that some children might be sensitive about their height.)

Encourage the children to use standard measures such as tape measures to calculate their height. Less able children may need adult support. For comparisons to be made, the same units need to be used.

Explain that the smaller feet will be sizes 10, 11, 12 and 13, whereas the larger feet will be sizes 1 and 2, etc.

Talk about/discuss:

Who is the tallest in the class?

Who has the largest feet in the class?

Are they the same person?

Health and growth

6. Do we get warmer or cooler when we exercise?

You will need: digital ear thermometer and/or Centigrade strip thermometers; stopwatches; copy of chart for each pair; plastic cups; water.

- In pairs the children should use the thermometers to take their temperature before they start exercising.
- Ensure they write down their start temperature. It may help to provide a chart like this for the children to complete.

Name	Start temp	After 1 minute	After 2 minutes	After water

- Tell the children that one person from each pair will jog whilst their partner times them. Explain they are jogging on the spot and not running up and down.
- After a minute, the children should take their temperature again. Ensure they record it.
- For the best results, it is a good idea to get the same child to jog on the spot again for another two minutes rather than swap with their partner at this point.
- After the two minutes, they should take their temperature again and note it down.
- Partners should then swap position and repeat the process.
- When they've finished, ask them to drink some cool water and take their temperatures again.

Talk about/discuss:

- ✦ Were they warmer or cooler after the exercise?
- ✦ Did their temperature increase with the amount of exercise they did?
- ✦ Did their temperature change after drinking the water?
- ✦ Explain that as we get older our bodies get better at maintaining temperature. When we exercise, our muscles make heat, which increases our temperature; our bodies then sweat, and as the sweat evaporates our bodies cool down again. Drinking water will help us to cool down.
- ✦ If we had used oral digital thermometers, how would drinking cold water have affected the readings?

Safety:

- ✦ If sharing thermometers, ensure the thermometer is cleaned before each child uses it.

7. Does exercise affect the rate at which we breathe?

You will need: PE kits; 10-minute timer.

- This activity is best conducted during a PE lesson. As with all experiments, ask the children to make a hypothesis first. In this case, do they think that exercise affects their breathing rate or not? Ask them to write down their prediction.
- Ask the children to find out their normal resting breathing rate by counting how many breaths they take in a minute (a breath is one inhalation and one exhalation).
- Encourage the children to generally work hard during the PE lesson.
- Use the timer and, after each 10-minute interval, ask the children to count how many breaths they take in a minute again. Also, make a note of what kind of exercise the children have been doing.

Talk about/discuss:

- Ask the children if their breathing got faster or slower.
- Back in the classroom, children can compare their findings with their original hypothesis.
- Explain that the reason their breathing increased is because the human body can only get oxygen through breathing, and the more exercise the body does, the more oxygen it needs.

Safety:

- Children should undertake normal PE activities, not attempt to 'test' their stamina or strength. Ensure that children who need inhalers have them readily available.

8. Does our pulse rate get faster or slower when we exercise?

You will need: PE kits; 10-minute timer; copy of chart.

As with Activity 7, this experiment is best conducted during a normal PE lesson. Ensure that the children are aware that they are testing something else and that they will need to make a separate hypothesis. Encourage the children to make their hypothesis and to write it down before they leave the classroom.

Before you start the PE lesson, ask the children to take their pulse. Some children may require help to find their pulse.

Ensure they write down their pulse rate. It may help to give each child a chart to complete similar to this one:

Name	Start pulse	After 10 mins	After 20 mins

Carry out a normal PE lesson and encourage the children to work hard as the timer approaches the 10-minute mark.

When the timer goes off, ask the children to take their pulse again and write it down.

Reset the timer and continue the lesson. When the timer is approaching 10 minutes, encourage the children to work hard again.

When the timer goes off, ask the children to take their pulse rate again.

If desired, you could also ask the children to take their pulse after the warm-down to compare the difference.

Talk about/discuss:

- ✦ Did their pulse rate get faster or slower after doing exercise?
- ✦ Tell the children that blood carries oxygen around the body. We can tell how fast our heart is beating from our pulse. Our pulse increases when we exercise because our muscles need more oxygen.
- ✦ Explain that the heart beats faster with exercise to get more oxygen to muscles in motion. The heart beats slower when we are resting because our muscles are less active.

Safety:

- ✦ Children should undertake normal PE activities, not attempt to 'test' their stamina or strength. Ensure that children who need inhalers have them readily available.

9. Does colour influence which food we like to eat?

You will need: graph paper; large paper; coloured pencils; a selection of different-coloured drinks (eg clear – cream soda and water; orange – carrot juice; red – tomato juice; cranberry juice and, if possible, blue and green fruit juices). Do not use food colourings.

✦ List the drinks available on the board. This experiment works better if more obscure drinks are chosen that the children may not have tried before.

✦ Demonstrate how to draw a bar chart on a large sheet of paper, listing the drinks along the bottom and the numbers up the side.

✦ Ask the children for a show of hands to discover which drink each would like to try first. Explain that they should choose only one, and complete the bar chart as you go. They need to taste only a tiny bit. Have water available for them to rinse their mouths if they really do not like a particular drink.

✦ Encourage them to tell you why they think this will be their favourite, eg look, smell, tasted before etc.

✦ Allocate each child a cup to taste the selection of drinks. Explain they need to choose one favourite.

✦ Draw another class bar chart and compare it to the first.

✦ Provide opportunities for the children to draw their own bar charts.

Talk about/discuss:

✦ What factors influenced their decision of which drink was their favourite?

- Explain that we use a lot of different factors to decide what foods we like, not just on the way it looks. We take information from our other senses (such as taste, smell etc) into consideration, too.

Safety:

- Do not forget to check if the children have any allergies, and get the necessary permissions before allowing the children to taste the drinks. Also, remember good hygiene is important at all times.

More ideas:

- The data collected could be used to construct pictograms using ICT.

Teeth and eating

10. Do all rabbits eat the same food?

You will need: paper; pencil.

- ✦ Explain that the children are going to plan a science investigation to attempt for homework.
- ✦ How could they find out about the diet of rabbits? Suggest books, Internet, their own pets. Discuss how they can find out if all rabbits eat the same food.
- ✦ Identify the different food that could be eaten by rabbits. Explain that some rabbits forage for food, so their food does not necessarily come from a packet.
- ✦ Which rabbits should be included in the investigation – wild, pets, or both?
- ✦ How would you describe the different sorts of food a rabbit eats, eg cereal, fruit, vegetables, grasses/plants etc?
- ✦ Help the children decide how they are going to collect the data and present the information. Suggest they use tables, pictograms or bar charts.
- ✦ Encourage them to conduct the investigation for homework by asking friends and family what their pet rabbits eat.

Talk about/discuss:

- ✦ Ask the children how good the evidence was.
- ✦ How many rabbits were in the sample?
- ✦ What, if any, conclusions can be drawn?

11. Do cats and dogs eat the same food?

You will need: pencil; paper.

This is a variation on Activity 10. Explain to the children that they are going to plan a science investigation to attempt for homework.

How could they find out about the diet of cats and dogs? Suggest books, Internet, their own pets. Discuss how they can find out if all dogs and cats eat the same food as each other.

Identify the different foods that could be eaten by dogs and cats. Explain that some cats hunt for their own food and that their food does not necessarily come from a tin.

How would you describe the different sorts of food dogs and cats eat, eg fish, chicken, rabbit, tinned food and dried food?

Help the children decide how they are going to collect the data and present the information. Suggest they use charts, pictograms or bar charts.

Encourage them to conduct the investigation for homework.

Talk about/discuss:

Ask the children how good the evidence was.

How many dogs and cats were in the sample?

What, if any, conclusions can be drawn?

12. Which foods are particularly damaging to teeth?

You will need: enough pennies for three per group; see-through plastic cups; selection of liquids, eg lemon juice, vinegar, sugar solution, salt solution, plain water, cola; small sticky labels.

- Split the class into small groups and give each group three plastic cups, each containing one of the three pennies.
- Let each group of children choose three of the liquids from the selection available.
- Pour a little of each of their chosen solution into each cup, and ensure the children label the cups clearly with the name of their group and the name of each solution.
- Leave the cups overnight. The following day, check to see if there is any change to the pennies. Continue to check the pennies each day for a week.
- Were there any changes? What happened to the pennies? What does this tell us about the particular foodstuff?

Talk about/discuss:

- Explain that although pennies are made of copper and teeth are covered in enamel, this experiment gives a good indication of why it is important to brush your teeth after eating, because your teeth are attacked in a similar way.
- Tell the children particular foods such as sweets are damaging to teeth, whereas other foods such as carrots and apples are less damaging.

More ideas:

- Compare different types and brands of cola, such as sugar-free and Zero, to observe if sweeteners also affect the pennies.

Moving and growing

13. Who have the longest arms – boys or girls?

You will need: Tape measures; graph paper; coloured crayons.

- ✦ In groups of four, two boys and two girls, tell the children they are going to find out who have the longest arms – boys or girls.
- ✦ Encourage the children to explain precisely which body measurement to take and how they are going to make the measurements so that reliable comparisons can be made.
- ✦ The children should present their results in the form of a table, bar chart or pictogram.
- ✦ Whose arms are the longest, the boys' or the girls'?
- ✦ Is this the same for each group?

Talk about/discuss:

- ✦ Allow time for each group to explain to the rest of the class what their tables, bar charts or pictograms show.

14. Are adults' heads bigger than children's heads?

You will need: selection of adults; selection of children; tape measures.

- Organize for four or five adult helpers who would be happy to have their heads measured to come to the class.
- Encourage the children to use tape measures accurately to measure the adults' heads.
- Ask for the same number of children to volunteer to have their heads measured.
- Ask different children to use tape measures accurately to measure the child volunteers' heads.
- Whose heads are the biggest – the children's or the adults'?

Talk about/discuss:

- Explain that the skeleton (including the skull) grows from birth to adulthood – approximately 21 years of age.

15. Do muscles work harder during exercise or when you are sitting still?

You will need: timer/stopwatches.

Ask the children to sit at their tables, place their elbows on the table and lean their head on their hands.

Gently squeeze the front of their upper arm, halfway between the shoulder and the elbow. Tell them that this is a muscle under their skin and that it is called a bicep.

What does their bicep feel like? Soft and relaxed, or hard and firm? Explain that the muscle is at rest and not doing any work.

Then, taking turns, they should put one hand under the edge of the table and carefully try to lift it a few centimetres from the floor.

Ask the children to feel their bicep with their other hand while they are holding up the table. What has happened to it now? Are their biceps soft and relaxed, or hard and firm?

Now feel the muscle in the lower half of the leg. This is called the calf.

Ask the children to jog on the spot for two minutes using the timer.

What do their calves feel like now?

Talk about/discuss:

✦ What does this tell you about your muscles?

- Explain that muscles help your bones to move.
- Reinforce that when a muscle is working it is firm, and when it is relaxed it is soft.
- Tell the children that when someone is exercising or moving fast, the muscles work hard. Explain that the muscle feels hotter because it has been working hard.

Keeping healthy

16. Are everybody's pulse rates the same?

You will need: stopwatches.

- Show the children how to find their pulse in their wrist.
- Divide the class into pairs. Ask the pairs to time each other's pulses for a minute. How many beats can they count in a minute?
- Now ask them to time for 15 seconds and multiply by four to get their average pulse rate.
- Are their pulse rates the same?

17. What sort of exercise affects pulse rate the most?

You will need: this will depend on the children's planned experiment, but have a selection of sports equipment available, such as skipping ropes, quoits, bean bags, small and large balls, cones etc.

- Explain to the children that they are going to plan and carry out their own science investigation to find out what sort of exercise affects pulse rate most.
- Ask them what forms of exercise they can think of.
- How do they think these exercises would affect their pulse rate?
- How could they measure it?
- How could they record their results?

Talk about/discuss:

- What did they feel like after exercising, eg tired, out of breath, hot?
- How did this compare to when they were sitting still?
- What happened to their pulse rate?

Safety:

- Explain to the children that they should undertake normal types of PE activities and not attempt to 'test' their stamina and strength. Ensure they have their inhalers readily available if they should need them.

18. Does your pulse rate stay higher for longer if you exercise longer?

You will need: this will depend on the children's planned experiment, but have a selection of sports equipment available, such as skipping ropes, quoits, bean bags, small and large balls, cones etc; stopwatches.

This experiment is a variation on Activity 17. Ask the children to adapt their previous experiment to find out if their pulse rate stays higher for longer if they exercise longer.

What variables would need to stay the same? What variables would need to change?

How could they make it a fairer test?

Talk about/discuss:

What happened to their pulse rate?

Was this what they thought would happen?

What have they learned?

Explain that our pulse rate increases whilst we are exercising because our muscles need more oxygen; when we stop exercising, our pulse rate returns to normal.

Growing plants

19. Do plants need soil to grow?

You will need: beans; jars/plastic cups; paper towel/kitchen roll; soil; water; sticky labels.

- ✦ Divide the class into small groups and give each group two plastic cups and four beans.
- ✦ Label their pots with their group name.
- ✦ Tell the children that in one of the cups they should put a little soil and in the other put kitchen paper or a paper towel.
- ✦ Ask them to place two beans in each pot.
- ✦ Supervise while they carefully water the beans and put the pots in a sunny place, such as a window sill.
- ✦ Remind them to water the beans a little every day.

Talk about/discuss:

- ✦ Encourage the children to use drawings to record their observations and to communicate what happens to the beans.
- ✦ Do they see any differences?
- ✦ What can they deduce from this?
- ✦ Reinforce that sometimes plants do not need soil to grow.
- ✦ Let the class observe what happens to the plants over time to discover which plants thrive and which ones don't.

20. Do plants need water to grow?

You will need: cress seeds; paper plates; paper towels/kitchen roll; sticky labels; water.

- Split the class into small groups. Give each group two paper plates and two paper towels.
- Ask the children to label their plates with their group name.
- Fold the paper towels and place one on each plate.
- Sprinkle a few cress seeds on each paper towel.
- Explain they are going to water one but not the other. It is very important they understand this, or the experiment will not work.
- Place both plates in a sunny place or window sill.
- After a few days, check to see if the seeds have grown.
- What can they notice?
- Why do they think this is?
- Reinforce the concept that plants do need water to grow.

21. Do plants need light to grow?

You will need: cress seeds; paper plates; paper towels; water.

This experiment can be carried out at the same time as the previous experiment. However, it is important at this level to explain that they are testing for two different things.

As a class, discuss which variables would stay the same and which variables would have to change to make it a fair test.

They will need one more paper plate per group. Ask the children to label their plates with their group name.

Remember that if you are not conducting this experiment at the same time as the previous one, you will need a control plate to place in a sunny place for comparison.

Fold the paper towels and place one on each plate.

Sprinkle a few cress seeds on each paper towel and water them.

Explain that they are going to put this plate in a dark place (such as a cupboard), and that they must remember to water both plates every day. It is very that important they understand this, or the experiment will not work.

After a few days, check to see if the seeds have grown.

What can they notice? Why do they think this is?

Explain that the cress kept in the dark is yellow because plants need light to be healthy. Also, increases in height occur because the plants grow towards the light if there are any cracks where the light can creep through.

Help them to record their findings.

22. Does the size of the pot influence the size of the plant?

You will need: 2 plants of the same type and size that are too big for their pots; 1 larger pot; more soil.

Show children the two pot-bound plants that have grown too large for their pots.

Take one plant out of its pot and repot it.

While doing this, show the children its roots.

Ask them why they think the plant needs repotting.

What do they think will happen to the plant now it is in the bigger pot?

What do they think will happen to the plant that did not get repotted?

Let the children observe the two plants over the next few weeks to see which one thrives.

Talk about/discuss:

What is happening to the plants?

✦ Encourage the children to sketch and label their observations of what happens to the plants over the next few weeks.

Plants and animals in the local environment

23. Where do living things live?

You will need: school playground/field or a park; magnifying glasses; copy of chart.

- ✦ Ensure there are flowerpots, stones and logs etc in suitable places a few days before this investigation takes place.

- ✦ Walk around the school or visit the local park to identify where plants are growing and where there are animals. Turn over stones and flowerpots to find woodlice, visit the school field to find daisies, explore under damp bushes and walls to find snails and dig up soil to find worms etc.

- ✦ Help children to make a brief record of what they found using a table such as the one below:

Location	Animal/plant	Location	Animal/plant
Field		Stones	
Wall		Logs	
Flo'pot		Trees	

Talk about/discuss:

- ✦ Discuss what animals and plants were found and where they were found.

- ✦ Was this what they expected?

Safety:

- ✦ When working outdoors, check that there is no broken glass, dog faeces etc. Ensure that the children wash their hands after handling the soil. Remember that all off-site visits must be carried out in accordance with LEA/school guidelines.

24. Which animals and plants live in different local habitats?

You will need: two contrasting areas, such as the playground, a playing field, an unpaved area under a tree, a school garden, a pond, a grassy area etc; magnifying glasses; large sheets of paper.

- Divide the class into small groups.
- Encourage the groups to make their predictions of what they might find in their allocated areas on large sheets of paper.
- Each group does not necessarily have to investigate the same two areas.
- Ask children to find out which animals and plants live in the two chosen areas.
- Help the children to record what they find by drawing and writing about the differences in their two allocated areas.

Talk about/discuss:

- Allow time for each group to report their findings to the class.
- What were the reasons for the differences?
- Did they find the plants and animals they expected?

Safety:

- When working outdoors, check there is no broken glass, dog faeces etc. Ensure that the children wash their hands after handling the soil. Remember that all off-site visits must be carried out in accordance with LEA/school guidelines.

Variation

25. Are all animals the same?

You will need: pictures of animals, including humans.

- ✦ Identify ways in which different types of animals are alike. Do they all have heads and eyes, do they all move etc?
- ✦ Identify ways the animals are different from each other, and classify into different categories, eg birds, fish, mammals, invertebrates etc.
- ✦ Ask the children to find other ways of classifying the animals, such as legs/no legs, fly/walk/slither etc.
- ✦ Ask the children to pick two animals from two different categories and list the similarities and differences in a chart.

Talk about/discuss:

- ✦ Identify ways in which humans are different from other animals, eg we walk on two legs/stand upright, we don't have fur, we have hair etc.

26. Are all humans the same?

You will need: photographs of each child.

- Ask the children to bring in photographs of themselves.

- Sort the photos into groups using their own criteria, such as boy/girl, eye colour, hair colour, shape of face, hairstyle, height etc.

- Split the class into pairs and ask each pair to write a description of their partner. If there are any twins in the class, it is useful to ask them if they have any differences.

- Be aware that some children may be sensitive about their appearance.

- Ask them to list similarities and differences.

Talk about/discuss:

- Explain that we are all unique.

- Discuss with the children how they can change their appearance.

- Is it possible for them to disguise themselves? Would they still be recognized?

27. Are all plants the same?

You will need: two differently structured types of flower, such as dandelion and buttercup or chrysanthemum and fuchsias.

- Split the class into small groups and give each group one of each flower.
- Ask them to draw both flowers.
- Encourage them to identify and label the parts that are common to both flowers, such as leaf, stem, petal, flower.
- Point out differences in shapes of leaf, colour of flower, thickness or woodiness of stem.

Talk about/discuss:

- What can they conclude from their observations?

28. Do people with the biggest feet have the biggest hand span?

You will need: centimetre rulers; sheets of A1 paper; shoe size information; marker pens; copy of chart.

Working in small groups, each group should copy the chart onto a large sheet of paper.

Name	Foot size	Hand size

The children should measure their foot and hand span with a centimetre ruler and jot it down on the chart.

Explain that they are all using centimetre rulers to make comparisons easier.

Discuss how to measure their hand span.

In each group, determine who has the biggest feet and who has the biggest hand span.

Are they the same person?

More ideas:

Using the children's charts, ask them to line up in order of who has the biggest feet and then in order of who has the biggest hand span.

Were their predictions correct?

✦ Explain that humans are similar to each other in some ways and different in others, and that some differences between themselves and other children can be measured.

29. Do people with the longest legs jump the farthest?

You will need: metre sticks marked in centimetres; large space to jump, such as the hall or field; cones or tape to indicate start position.

Determine who has the longest legs by lining the children up in order of leg length. At this level, an approximation should be made simply by making an observed judgement on who has the longest legs.

In small groups, the children should run and stride jump from a determined point.

Who jumped the farthest?

Was it the person with the longest legs?

Talk about/discuss:

Ask how they could have made it a fairer test.

Suggest they could have each taken their own inside leg measurement before they started, rather than guessing.

Safety:

I do not recommend taking inside leg measurements, and suggest you use the fact you did not as a teaching point on whether it was a fair test.

30. Do people with the longest arms throw the farthest?

You will need: metre sticks or trundle wheels; tape measures; large space to throw, such as the field or playground; large balls; cones or tape to indicate start position.

✦ In groups of three, ask the children to measure the length of their arms from the point of their shoulder to the tips of their fingers. Explain they should all measure in centimetres to ensure it is a fair test.

✦ This experiment is best conducted outside. Use large balls, as they can't throw them as far and so it is easier for them to measure the distance. Explain that they are going to measure from the first bounce rather than from where the ball stops.

✦ Each child in the group should take it in turns to throw, while one watches for the first bounce and stands on this point; the other person can then measure using a metre stick or trundle wheel the distance between the two children.

Talk about/discuss:

✦ Who threw the farthest in the group?

✦ Did they have the longest arms?

✦ Explain that there are lots of other variables that should be taken into consideration (eg, muscle strength and physical fitness), and not just arm length.

Helping plants grow well

31. Do plants need leaves to grow?

You will need: two healthy plants of the same size (geraniums are a good choice) water; rulers; chart.

- Show the children two similar plants of the same species.
- Ask the children to suggest ways these plants could be used to investigate if plants need leaves to grow.
- Respond to the children's suggestions.
- Conclude that they could remove the leaves from one plant.
- Ensure that the children realize that they should keep the plants in the same place and give them the same amount of water.
- Over the next few weeks, encourage the class to carefully record the height and diameter of the plants.

Talk about/discuss:

- Display their measurements on a chart.
- What can they conclude from this experiment?

32. How do plants feed?

You will need: white flowers, such as carnations; celery; two vases; plastic cups; food colouring; hour timer; water.

Show the children a complete head of celery and ask them to look closely at the individual stalks of the plant.

Cut a stalk width-wise, then observe the tiny holes in the cut end.

Put the celery upright in a shallow container of water with a few drops of food colouring added.

Check the celery every hour.

What happens to the celery?

Ask the children to make drawings to show what they observe and annotate what they think has happened.

Now ask the children what they think will happen to the white flowers if they are put in the food dye.

Pour some water and several drops of food dye into a vase and add half the flowers; put the rest in a vase of plain water.

Over the next few days, observe and record the differences.

Talk about/discuss:

Explain that the water containing the food dye has travelled up through the veins of the celery and the flowers and that that is the reason why they change colour.

✦ Tell the children that plants need water to survive.

33. Do plants grow better the more they are watered?

You will need: jug; beans that have begun to sprout; water; rulers.

- ✦ Look at and discuss the seedlings and what they need to grow.
- ✦ Ask what evidence the children would need to collect to find out if the plants grow better the more they are watered.
- ✦ Measure specified amounts of water.
- ✦ Suggest they give one of the seedlings no water, one 10ml and one 100ml every other day for a few weeks.
- ✦ Observe and record the effects on the height of the seedlings.

Talk about/discuss:

- ✦ Explain that plants need water to grow, but too much water will kill them.

34. Does gravity affect the way plants grow?

You will need: light-proof box; pots of fully grown cress or mustard seeds; Plasticine®.

- Place the pots of cress or mustard seeds into the box. One should be laid on its side, and can be secured with the Plasticine® so it does not roll around; the other should be placed upright as a control pot.

- Ensure the cress (or mustard seeds) are watered, and then put the lid on the box.

- Check the box each day to see what has happened to the cress or seeds.

Talk about/discuss:

- Explain that the cress or mustard in the pot on its side has grown upwards because the shoots are sensitive to gravity. Seeds are usually planted beneath the soil and will grow up towards the surface away from the downward force of gravity.

- Tell the children that the control pot was used to check that the cress or mustard was not changing direction because light was getting into the box.

Habitats

35. How do we know woodlice prefer damp conditions?

You will need: woodlice; glass tank; logs, leaves and stones; soil or sand; water.

- Collect some woodlice.
- Set up different areas in the tank using stones, leaves, wood. Put the same on the opposite side of the tank.
- Ensure on one side the soil is damp (not soaking wet); on the other side of the tank, the soil should be dry.
- Make and justify a prediction, such as 'The woodlice will be under the stones on the damp side, because they prefer to be where it's dark and damp.'
- Observe which side of the tank the woodlice prefer.

Talk about/discuss:

- Encourage the children to describe the habitat in terms of the conditions. For example, leaf litter is cool, damp and dark.
- Explain that not all animals and plants are found in the same places, because different types of animal/plant like different conditions.
- This information could be collated in an ICT data-handling program.

Safety:

- If animals are brought into the classroom, ensure that they are treated sensitively and their needs met and that they are returned to the habitat from which they came as soon as possible. Children should wash their hands after handling animals.

36. How do we know mealworms prefer the dark?

You will need: containers; torches; card; mealworms (it is possible to purchase mealworms from pet shops and fishing shops); timers.

Split the class into small groups.

Each group will need a container, a torch, some card big enough to cover the top half of the container and some healthy mealworms.

Discuss how many mealworms they should use and distribute them accordingly.

Ask the children to put the mealworms in the container and cover half of it with the piece of card.

Shine the torch directly down over the open side of the container. Be careful not to shine the torch through or under the card.

Discuss how long the mealworms should be left and agree a set number of timed intervals to observe them (eg 5 mins, 10 mins and 15 mins).

Use the timers and observe which side of the container the mealworms are in at each of the timed intervals.

Talk about/discuss

- Did the mealworms move toward the light or the dark side?
- Explain that the mealworms usually live underground.
- If you transfer the mealworms to a jam jar half-filled with a mix of flour and bran cereal, it is possible to observe the whole lifecycle process. Feed regularly on pieces of apple or potato.

37. Do earthworms live above or below the ground?

You will need: worms; soil, sand and fine gravel; jam jars; 1 bucket per class; spades; large area where the children can look for worms; water.

✦ Discuss with the children where they expect worms to live and how they might collect some for their experiment.

List what equipment they think they will need.

In small groups, the children can hunt for worms.

With the worms that they find, help them set up wormeries in jam jars with a mix of soil, sand and fine gravel. Ask the children to predict what the worms will do when they are put in.

Encourage them to make careful observations.

Talk about/discuss:

Where did they find their worms?

How did they encourage the worms to come to the surface?

What happened to the worms when they were put in the wormeries?

Discuss the children's observations and ask them to explain these in terms of what they already know about worms and their habitats.

Safety:

✦ If animals are brought into the classroom, ensure that they are treated sensitively and their needs met and that they are returned to the habitat from which they came as soon as possible. Children should wash their hands after handling animals.

38. What do snails like to eat?

You will need: snails; large container; lettuce; porridge oats; dandelions; sunflower seeds.

- Ask the class how we can find out what snails prefer to eat and list their ideas.
- Show them the different food types you have available for the experiment.
- Ask them to predict whether they think the snails would like each of the food types provided.
- In small groups, place one type of food in each corner of a container.
- Put the snails in the middle.
- Observe which food they go to eat.
- After a few minutes, remove the snails from the food and place them back into the middle of the container.
- Re-release the snails and see if they go back to the same food type.
- Repeat this a few more times to ensure it was not just coincidence.

Talk about/discuss

- Which food type did the snails prefer?
- Was this what the class had predicted?
- Ask how many snails were used. Discuss if this was a large enough sample to investigate.

Life cycles

39. How many ways can seeds be dispersed?

You will need: collection of different types of seeds (eg grasses, winged seeds, fruit) and/or pictures of various seeds, including some unfamiliar seeds; seed packets; pictures of the parent plants of the seeds being investigated; magnifying glasses.

- Offer children a range of seed packets and seeds to examine.
- Talk about the differences between the seeds.
- Use magnifying glasses and talk about their observations.
- Make observational drawings.
- Talk about seed dispersal and encourage the children to suggest how they think the seeds are dispersed.
- List the children's suggestions and try to classify the seeds into broad groups, depending on how they are dispersed (wind, animals, water etc).
- Look at unfamiliar seeds. Ask the children to suggest to which group each of the seeds belong.
- Use books and the Internet to check whether their ideas were correct.

Talk about/discuss:

- Explain that seeds can be dispersed via four ways: water (eg coconuts), wind (eg dandelions, maples), explosion (eg gorse, broom, yellow balsam) and animals' digestion (eg grasses, acorns, blackberries).

✦ Tell the children that seeds need to be dispersed so they have the best chance of growing into a new plant.

Safety:

✦ Avoid using nuts because of allergies. Do not let the children taste any of the seeds.

40. Which seeds have the most effective parachutes?

What you need: seeds and fruit with parachutes or wings (eg, dandelions, sycamore and ash seeds); tweezers; magnifying glasses; container to keep the seeds in when they are not in use.

Predict which seed has the best parachute.

This experiment is better conducted indoors, as the wind can have an adverse effect on the results.

Working in pairs, the children should have races with the different seeds to see which falls the fastest and which falls the slowest.

Use tweezers to lift the seeds so as not to damage their parachutes and wings.

Explain to the class that the slower they fall, the better the parachute. The parachutes are meant to help disperse the seeds, and a seed's weight helps determine the type of parachute (downy parachutes for tiny seeds, propellers for heavier seeds etc).

Did the children accurately predict which seed had the best parachute?

Talk about/discuss:

Ask what they observed.

What factors might help to determine the size and shape of a seed's parachute?

Encourage the children to suggest reasons why some seeds might not grow into plants.

41. What do seeds need to germinate?

You will need: radish seeds; flowerpot for each child; bag of compost; stones for drainage; water; a dark place to put pots; a sunny place to put pots.

- Divide the class into groups, so there are equal numbers of the following pots:
  - Control pots that will have water, sun and soil.
  - Dark pots that will have water and soil but no light.
  - Dry pots that will have soil and light but no water.
  - Soil-free pots that will have water and light but no soil.

- Discuss how they are going to make the investigation a fair test.

- Each group should plant their radish seeds in their pots, label them and position them in the relevant place.

- Remind the children to check their pots each day and tend to them appropriately, keeping in mind the conditions they have to adhere to.

- What do the children observe?

- Are some plants growing better than others?

- What conditions did the seeds need to germinate?

More ideas:

- Encourage the children to write up their experiments, showing their prediction, equipment used, method, results and conclusions.

Enquiry in environmental and technological contexts

42. Does location affect a dandelion's growth?

You will need: sheets of large paper; marker pens; a wild area that has plenty of dandelions to observe (this experiment could be carried out with other plants, such as bluebells, snowdrops, buttercups, daisies etc); rulers; ICT spreadsheet software.

- Split the class into small groups and ask them to decide how they could investigate what affects the growth of a dandelion.
- Ask them to note on sheets of A1 paper:
  - What areas of the dandelion they would use to decide how well it grows.
  - How they will collect the evidence.
  - What they will need to measure and when.
  - What size sample they will need to use.
- Encourage them to identify questions for investigation. For example, does the height of the grass around the dandelion affect the width of the stems? Does a dandelion under a tree grow as well as a dandelion in direct sunlight? Do dandelions in short grass have larger flowers?
- Provide opportunities for the children to carry out their investigations and collect and record data accurately.

More ideas:

- Enter the data they have collected into an ICT database and encourage them to analyse the data by producing graphs and bar charts.
- Help the children to suggest reasons for any differences found in the dandelions.

43. Do you find more small insects on the top or underside of leaves?

You will need: selection of leafy shrubs; magnifying glasses' large sheets of paper; marker pens; ICT spreadsheet software.

- Split the class into small groups and ask them to decide how they could investigate whether small insects prefer the top or the underside of leaves.

- Ask them to note their group ideas on the A1 paper.

- They will need to consider:
  - How many leaves they would need to check.
  - How many different trees and plants they would need to check.
  - How they will record how many insects they see on each leaf and where they are situated on the leaf.
  - How they can conduct the investigation without hurting the insects or the plants.

- Encourage the children to identify questions for investigation such as:
  - Does the size of the leaf affect which side the insects prefer?
  - Does the weather affect the insects' preferences?
  - Does the bushiness of a shrub affect where you find insects?

- Provide opportunities for the children to carry out their investigations and collect and record data accurately.

- Ask the children to collect data and make observations where appropriate.

- Identify patterns in the data and help the children look critically at results to decide how strongly they show a trend, particularly in relation to sample size.

Talk about/discuss:

- Help the children to suggest reasons for their findings.
- Use an ICT data-handling spreadsheet to collate the data and make graphs of their results.

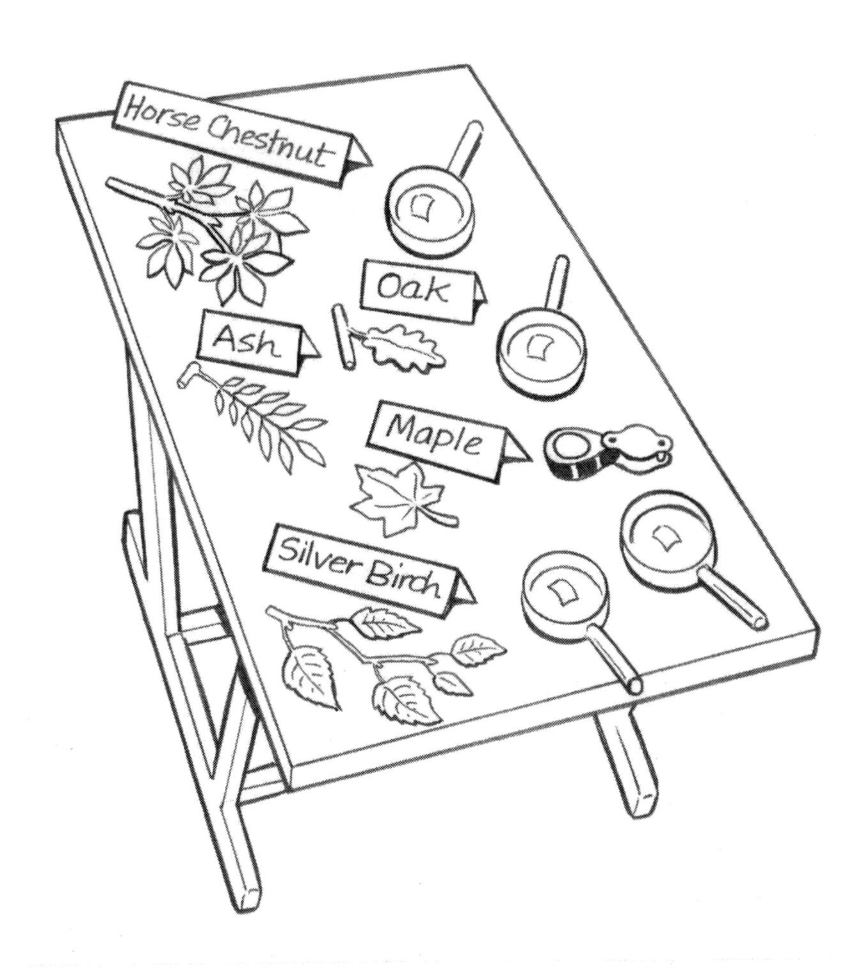

44. Is there any difference in the length and width of leaves on the top, middle or bottom of a shrub?

You will need: variety of shrubs; rulers; ICT spreadsheet software.

- Ask the children to plan with a partner how they could develop a fair test to find out if there is any difference in the length and width of leaves on the top, middle or bottom of a shrub.
- Encourage them to think about:
  - The size of the sample they should use.
  - How many leaves should they measure.
  - How many leaves they should use from the top, middle and bottom of the shrub.
  - Whether they need to investigate more than one shrub.
  - Whether they should use the same species of shrub or a variety.
- Explain that they need to be careful that they do not damage the plants.
- The children need to decide how they are going to record their findings.
- Provide an opportunity for them to collate their data using ICT.

Talk about/discuss:

- Are there any patterns in the data?
- Encourage them to describe the limitations of their evidence.

45. Will we find different animals at different depths of a pond?

You will need: nets; trays; pipettes; magnifying glasses; pond; identification charts; ICT spreadsheet software.

Encourage the children to work together in small groups to decide how they can create a fair test to find out if there are different creatures at different depths of a pond.

What evidence will they need to collect?

Ensure they formulate a hypothesis.

Discuss how many times they will need to dip the pond and at what depths to produce an adequate sample.

How will they measure the depth to make sure they can accurately dip there again?

How will they record the results (eg drawings, charts, photographs)?

Use the classification charts to identify the pond creatures the children find. This will help them to discover whether different animals are found at different depths of a pond.

Safety:

Explain to the children how to stay safe near ponds. Remind them not to lean too far over the edge where the water may be deep. Also, ensure they wash their hands after touching the pond water.

More ideas:

Compare each group's findings to see if any patterns emerge.

Use ICT database software to analyse the data in further detail.

Interdependence and adaptation

46. How do fertilizers affect plant growth?

You will need: three pot plants; liquid plant food; water; jugs; a sunny window ledge.

- Place all three plants in a sunny location and label 1, 2 and 3.
- Arrange for monitors to regularly water the plants with equal amounts of water.
- Explain that plant 1 will have plain water, plant 2 will have a few drops of the liquid plant food added to its water (as recommended on the container) and that plant 3 will have triple the amount of plant food added to its water.
- Over the next few weeks, observe whether there is any difference in growth in the three plants.
- This experiment could take up to seven weeks.

Talk about/discuss:

- Look at the packaging of fertilizers and ask the children to suggest why plant food is needed.
- Explain that plants take nutrients, as well as water, through their roots. Too much fertilizer has a detrimental effect on the plant, as only small quantities are needed.

47. What happens to plants grown in the dark then placed on a window sill?

You will need: 2 pot plants (you may be able to use the plants used in the previous experiment); a dark place; a sunny place.

- ✦ Review what the children know about what plants need to grow.
- ✦ Ask the children to suggest what will happen to a plant if it is placed in the dark.
- ✦ Put one of the pot plants in the dark but carry on watering it as before.
- ✦ What happens to the plant after two weeks?
- ✦ Put the plant back on the window sill and observe if the plant becomes sturdier and develops more leaves.
- ✦ Ask the children to make relevant observations and sketches to provide information about how the plant is growing.

Talk about/discuss:

- ✦ Explain that green plants need light in order to grow well.

Micro-organisms

48. How can you prevent food becoming mouldy?

You will need: apples; sealable bags; water; vinegar; mud; salt; sugar; fridge.

- Cut the apples into halves.
- Working in small groups, ask the children to half-fill sealable bags with either vinegar, water, mud, salt or sugar. Make sure you have enough groups to cover all the scenarios.
- Place half an apple in each bag and seal them.
- Put the bags in a cool place somewhere in the classroom.
- Have a control bag with nothing but the apple in it; place a similar control bag in a fridge.
- Encourage the children to observe and to record their observations of all seven bags for three weeks.

Talk about/discuss:

- Which apple went mouldy the quickest?
- Which apple was the best preserved?
- How can you prevent food from becoming mouldy?

49. What does yeast need to stay alive?

You will need: packets of active dry yeast; warm water (approx 45°C); plastic cups; tablespoons; thermometers; sugar; large rubber balloons; empty half-litre plastic bottles; marker pens.

This experiment can be teacher demonstrated, or children could work in small groups. It may be necessary to organize extra adult supervision.

Stretch out the balloon by blowing it up and deflating it repeatedly, and then set it aside.

Place the balloon with the neck end towards you and draw a funny face on it.

Add a packet of yeast and two tablespoons of sugar to the cup of warm water (approx 45°C) and stir.

When the yeast and sugar have dissolved, pour the mixture into the plastic water bottle. The water will start to bubble as the yeast produces carbon dioxide.

Stretch the balloon neck over the mouth of the bottle and place it somewhere warm, such as near a radiator.

After several minutes, the balloon will stand upright. If you don't see anything happening, keep waiting. Eventually, the balloon will inflate and the face should be grinning at the class.

Try the same experiment using hotter and colder water. Use thermometers to measure the temperature of the water.

At what temperature is the yeast most active? Would the experiment work without sugar?

- At what temperature is it not possible to blow up the balloon?

Talk about/discuss:

- Explain that the rate the balloon inflates is proportional to the growth of the yeast. As the yeast population grows, so does the amount of CO^2 it produces and this is why the balloon inflates.

- Explain that yeast cells die at temperatures greater than 55°C.

- Make bread and explain how this process makes the bread rise.

Safety:

- Attention needs to be given to hygiene if children are blowing up the balloons. Do not let the children eat or taste the yeast.

Sorting and using materials

50. Which material would be best to make a toy slide?

You will need: selection of materials for children to examine, such as tinfoil, bubble wrap, cardboard, wood, linoleum etc.

✦ Tell the children you want to find a material that is suitable for making a toy slide.

✦ Ask the children to suggest what the material should be like and list their suggestions.

✦ Encourage the children to identify the properties of the different materials, such as rough/smooth, bendy/not bendy etc, and place the materials available into sets.

More ideas:

✦ As a design-and-make activity, encourage the children to make a slide using some of the materials they identified.

51. What materials stick to a magnet?

You will need: selection of different materials, such as paper, tinfoil, cork, sponge, wood, aluminium etc; magnets; copy of chart.

- Present the materials to the children.
- Ensure they are familiar with the names of the different materials.
- Ask them to predict which materials will be attracted to the magnet.
- Allow time for them to test their predictions.
- Encourage the children to make a record of what they find out.
- Use a tick chart similar to the one below:

Material	Magnetic:	
	Yes	No

Talk about/discuss:

- What did they observe?
- Explain that some materials are magnetic, but most are not.
- Which material(s) did they find was magnetic?
- Can they suggest any other materials that might be magnetic?

52. Which paper is best for wrapping a present?

You will need: selection of different types of paper, eg shiny, thick, rough, patterned, tracing paper, tissue paper, writing paper, newspaper, brown paper etc.; boxes to wrap; Sellotape®; scissors.

- Look at the different types of paper and encourage the children to describe them.

- Discuss what features would be ideal for wrapping a present, such as strong, easy to write on, easy to fold, opaque etc.

- Encourage the children to predict which paper out of the selection available would be best for wrapping a present as a gift or wrapping a present to post.

- Allow time for the children to test their ideas.

Talk about/discuss:

- Which paper was the best?

- How did they find out?

53. Which material would be best to make an umbrella?

You will need: selection of materials cut into squares of approximately $10cm^2$ – drawing paper, paper towel, newspaper, kitchen roll, bin liner, tinfoil, tissue paper, fabric, felt, fur (enough for one sample per pair); pipettes; plastic cups; elastic bands; water.

- ✦ Give the children the selection of materials and tell them they need to test each material to see if it would be any good to make an umbrella.
- ✦ Ask them what an umbrella would need to be like (encourage them to suggest 'waterproof').
- ✦ Tell them that they are going to test the different materials available to see if they are waterproof.
- ✦ Explain that they will need to secure the material to the cup and slowly drip the same amount of water into the middle of each piece of material with a pipette. If there is water in the cup afterwards, the material is not waterproof.
- ✦ Group children into pairs. Allow time for each pair to test all the materials available. Put them in the order of most waterproof, using 1 as most waterproof and 10 as the least.

Talk about/discuss:

- ✦ Help the class to discuss methods and results. Did they notice that some of the materials absorbed the water and got wet?
- ✦ Which material did they find to be the best?

More ideas:

- ✦ As a design-and-make task, encourage the children to make an umbrella with the material they found to be the best. The umbrella does not need to shut.

54. Which type of ball is the bounciest?

You will need: a selection of balls, such as tennis balls, basketballs, footballs, golf balls, ping-pong balls and cricket balls.

- Group the children into pairs and ask them to design an experiment to see which ball bounces the highest.
- Give them at least four different types of balls to test.
- Discuss with the children how to design a fair test. Suggest that they drop the ball and see how high it goes, rather than throw it. They must start at the same point each time to ensure it is a fair test.
- Encourage them to record their findings as they go along.

Talk about/discuss:

- Allow time at the end for each group to report their findings to the class.
- Which was the bounciest ball?
- Why do they think it was the bounciest?

Grouping and changing materials

55. What happens to shape and volume when liquid is poured from one container to another?

You will need: a selection of different-sized containers and jugs; 1 litre measuring jugs; water; wet play area.

- Present the children with the different-shaped containers.
- Ask the children to find out and record on a chart of their own design what happens to shape and volume when liquids are poured from one shaped container into a different-shaped container.
- Measure 1 litre of water and pour the water carefully into different containers.
- Make careful observations and measurements of volume to draw conclusions and make generalizations.

Talk about/discuss:

- Explain that liquids do not change in volume when they are poured into a different container.

More ideas:

- This experiment can also be used for revision when looking at solids, liquids and how they can be separated.

56. How can we make ice melt more quickly?

You will need: water; ice cubes of the same size; plastic cups; sunny window; radiator; timer; copy of chart.

- ✦ Demonstrate how an ice cube melts if left out of the freezer.
- ✦ Ask the children to touch the ice, describe what it feels like and observe what happens to its shape.
- ✦ Ask the children how they think they could make the ice cube melt more quickly.
- ✦ List their ideas and establish that the ice cube will melt quicker in a warmer place.
- ✦ Ask how they could make it a fair test. Explain that the ice cubes all need to be about the same size.
- ✦ Place five ice cubes in different positions around the classroom, such as on a radiator and on a sunny window sill and in several cooler places, and observe them every 15 minutes.
- ✦ Encourage the children to fill in a chart like the one below to record their observations.

Place of ice cube	Time observed	What we observed

Talk about/discuss:

- ✦ Discuss their observations and help them to draw conclusions from their work, such as ice melted faster when placed on a radiator, or the window sill was the warmest place, therefore, the ice melted more quickly.

57. What happens to liquids when they are cooled?

You will need: a selection of liquids, such as soup, water, tomato sauce, orange juice, yogurt, milk; ice cubes; ice-cube box (or ice-cube trays); freezer.

- Ask the children how they could turn water into ice.
- Show them some ice cubes and explain that before they went into the freezer, they were liquid (water).
- Ask the children to suggest what other liquids might change when they are cooled.
- Use a selection of liquids and freeze them.
- Encourage the children to make observational drawings of what has happened to these liquids.

Talk about/discuss:

- Explain how the liquids have changed. Use words such as hard; expanded; cold; bigger; solid. Make a list.
- Discuss whether this observation was what they expected.

58. How can we make chocolate softer?

You will need: chocolate; metal spoon; nightlight candle; matches.

Ask the children to suggest how they could make chocolate softer and list their ideas.

Test their ideas.

Melt a small sample of chocolate using a nightlight candle by placing it in a metal spoon and holding the spoon over the flame.

Was the result what they were expecting?

Talk about/discuss:

Ask the children to predict what will happen if they leave the material to cool.

Explain that materials, such as chocolate and butter become hard when they are cooled and become soft when they are warmed again.

Safety:

Children should be kept away from naked flame.

59. What happens when water is boiled?

You will need: water; measuring jug; kettle.

- Measure 1 litre of water and pour it into a kettle.
- Boil the kettle.
- What do the children observe when it comes to the boil?
- Point out the steam.
- After the kettle has boiled and the water has cooled slightly, pour the boiled water back into the measuring jug.

Talk about/discuss:

- How much water is left?
- What has happened to the rest of the water?
- Explain that the steam they saw was the water evaporating.

Safety:

- Children should be kept away from boiling water.

Characteristics of materials

60. What material makes the best floors for different rooms in a house?

You will need: selection of materials, such as linoleum, wood etc; large box; objects for dropping.

Explain that different types of materials are used for making different things, and that some materials are more suitable for making particular items than others. This is because each material has different characteristics or properties.

Discuss which materials they think would be best for a floor in a kitchen, compared to a living room or a garage.

List their ideas.

Why have they chosen these materials?

List their properties, such as strong, hard and flexible. A chart could be used like the one below:

Room	Materials	Properties

Ask the children how they would find out whether a material was hard and how they could compare the hardness of materials in order to find out what materials would be suitable for a floor covering for a model house.

Ask children to explore different ideas, such as rubbing or dropping things on the material.

✦ How could they make it a fair test?

Safety:

✦ Care is needed if heavy objects are dropped. Remind the children to keep their feet out of the way. A good idea would be to place the test material inside a large box before dropping objects on top. Do not use materials that might shatter or splinter, such as tiles.

More ideas:

✦ Encourage the children to explain whether they carried out a fair test or not and what made it fair or what made it an unfair test.

✦ Ask them to write sentences to explain their observations, such as 'The … would make the best floor in a (name of room) because … '.

61. Which type of paper is best for mopping up spills?

You will need: three different types of paper, such as kitchen roll, school paper towels and newspaper; water; syringes; trays; measuring jugs; magnifying glasses; copy of chart.

- Show the children the three types of paper and ask which they think will be the best for mopping up spills and why.
- Explain that they are going to plan a fair test to see if they are right.
- Tell the children to use the trays to contain the spills so the water does not go everywhere.
- Demonstrate ideas of what an unfair test would be, such as different amounts of paper towel, different amounts of water.
- Ask what equipment they should use to measure water.
- Look at the jugs and syringes and ask the children to consider how much water to use each time.
- Decide if the water should be syringed onto the paper, or if the paper should be laid on top of the water spill.
- Encourage the children to measure the water carefully, and to record their observations in a chart similar to the one below:

Type of paper	Amount of water absorbed	Ranking
Kitchen roll		
School paper towel		
Newspaper		

62. Which supermarket has the strongest carrier bags?

You will need: a selection of carrier bags from at least three different supermarkets; two tables; broom handle or strong dowelling; weights.

- ✦ Split the class into small groups. Each group should have at least three different supermarket shopping bags to test.
- ✦ Suspend all the carrier bags from the same broom handle, or dowelling, balanced between two tables.
- ✦ Add the weights to each bag evenly until either the bag's handles break or the bottom gives out.
- ✦ Count the weights in each bag.

Talk about/discuss:

- ✦ Which supermarket bag held the most weight?
- ✦ Was it a fair test and why?
- ✦ Discuss how, if the bag had been used previously, it may have affected the results.
- ✦ Did all the groups get the same results?
- ✦ Discuss with the children how some plastic bags can be recycled and the importance of recycling to the environment.

Safety:

- ✦ Remind the children to keep their feet out of the way when the bag breaks. Also, remind the children not to put the plastic bags over their heads, as this can cause suffocation.

63. How far can an elastic band stretch before it breaks?

You will need: a range of different-sized elastic bands; sets of 250g hanging weights; two tables; broom handle or strong dowelling; 30cm rulers.

Discuss any patterns the children might predict regarding the stretch and what possible load the elastic bands might withstand.

Encourage them to consider: What sort of evidence would support their predictions? What variables would they need to measure? How could they record their data systematically?

Show them the equipment available. In small groups, they should discuss how they could set up an experiment to test their predictions.

Suspend the elastic bands from the same broom handle or dowelling, and balance between two tables.

Add the weights, one at a time, to the elastic bands and keep a total of how much weight has been added. Take measurements of the 'stretch' of the elastic bands.

How heavy a load did the elastic bands take?

Talk about/discuss:

Did all the elastic bands stretch the same amount?

Was there a pattern to what happened to the elastic band as you increased the load? If you put on twice as much weight, did you get twice as much stretch?

Safety:

Remind the children that when elastic bands break, they can ping into the air. Because of this, they should stand well back and also keep their feet out of the way of the falling weights.

64. What material would make the best swimming costume?

You will need: A1 paper; marker pens.

- ✦ Split the class into small groups of about four children each. Explain that they are going to plan an investigation but will not be conducting it.
- ✦ Ask them to think about a new material that has been recommended to make a swimming costume.
- ✦ Ask the children to suggest what tests they would carry out on this new material to compare it with a familiar material already used to make swimming costumes.
- ✦ Explain that they will need to consider stretchiness and comfort.
- ✦ The children should write their ideas and plan the experiment on the A1 paper, listing the equipment they would need and what they would do (their method).
- ✦ How would they ensure that the comparison between the two types of material would be fair?

More ideas:

- ✦ Encourage each group to explain their plans to the class and to ask questions and comment on each other's ideas.

Rocks and soils

65. Which types of rock are the hardest?

You will need: a selection of different rocks, including chalk, clay, granite, marble, pumice, sandstone and slate (ensure there is enough of each type of rock for the children to experiment on); copper coins; pictures of granite worktops, marble tiles, ornaments, fireplaces, pumice stones and slate roofs; copy of chart.

✦ Examine the rocks and ask the children how they look and feel. Are they rough or smooth? Light or heavy? Hard or soft? Shiny or dull? Colourful or plain?

✦ Use a chart like the one below for the children to record their observations.

✦ Rub each rock gently with a copper coin to see how easily they are worn away.

✦ Rank the rocks in order of how hard they are, with 1 being the hardest.

Rock type	Observations	Uses	Properties	Rank

Safety:

Rocks should be rubbed gently and care taken to make sure the particles do not get into children's eyes.

More ideas:

Explain that different types of rock are used for different purposes, depending on their characteristics, and show pictures of how they can be used.

66. Which rocks are permeable?

You will need: a selection of different rocks, including chalk, clay, granite, marble, pumice, sandstone and slate – ensure there is enough of each type of rock for the children to experiment with; pipettes; water.

- Split the class into small groups and give each group a selection of rocks to investigate.
- Tell the children they should slowly drip small quantities of water onto the rocks and observe whether it remains on the surface of the rock or not.
- Do any of the rocks absorb the water?
- Explain that if the rocks absorb water, it means the rock is permeable. Impermeable rock does not absorb water.

More ideas:

- Explain that impermeable rocks have many uses because of their characteristics; hardness, strength and inflexibility.
- Encourage the children to compose scientific sentences such as, 'A roof is made of slate because it is impermeable and will protect things from the rain.'

67. Can you change the speed that water runs through soil?

You will need: clear plastic cups; elastic bands; school paper towels (not kitchen roll); soil; water; sand; jugs; labels; buckets.

Label the cups:

Soil
Soil and sand
Water (need two cups of water with the same amount of water in each)

Lay a paper towel over the top of each of the soil and the soil and sand cups. Push the towels about halfway into the cup and fasten with an elastic band.

Mix together $\frac{1}{4}$ cup of soil and $\frac{1}{4}$ cup of sand and carefully pour it into the paper towel in the soil and sand cup.

Carefully pour $\frac{1}{2}$ cup of soil into the paper towel in the soil cup.

Measure about 2cms of water into each of the plastic cups.

The children should pour the water into both the soil cup and the soil and sand cup at the same time.

Observe how much water passes through the paper towels and into the cup.

Ask the children what they think will happen if more water is added to each sample. Test to see if they were correct.

Talk about/discuss:

- Did all the water drain through into both plastic cups?
- Did the water drain through at the same speed?
- Explain that soil absorbs water, but by adding sand, the water will drain more easily through the soil. If a plant's roots get too wet, the plant will die, but if the drainage is better, the roots will be less wet and will not rot.
- Discuss with the children other methods of improving drainage, such as adding stones or gravel.

Keeping things warm or cool

68. How can we stop an ice cube from melting?

You will need: selection of materials, including newspaper, cotton wool, bubble wrap, felt etc; plastic cups; ice cubes; cardboard; timers.

- Explain to the class that they are going to plan their own investigation and carry it out.
- Divide the children into pairs. Allow the children time to create their own fair test to see which material is best at keeping things cold.
- The children should be allowed to experiment.
- Some children may wrap up the ice cube and some may use the plastic cups and insulate the cups. If they do the latter, it may be an idea to suggest that they make a lid with the cardboard and insulate that, too.
- Remind them that, for it to be a fair test, they must keep the variables the same except for the one they are testing, which is the material.

Talk about/discuss:

- Allow time at the end for the children to report back to the class about what they did and what they found out.
- How could they use this knowledge?
- Could the same materials be used to keep a drink or their sandwiches cool?

69. How could you keep a drink cool?

You will need: ice cubes; small bottle of chilled water per group; plastic cups large enough to fit the small bottle in; selection of materials, including bubble wrap, cotton wool, fur fabric, felt, bin liner, tinfoil, newspaper etc; timers.

A continuation of the previous investigation demonstrating how scientific findings can be put to practical use.

Ask the children for ideas on how they could keep a small bottle of water cool.

Split the class into small groups and give a small, cold bottle of water to each group.

Encourage them to write down what they would do and what apparatus they would use. Would they use the same material and ideas that they used for the ice cube? Why? Or why not?

Allow time for them to experiment.

Some may wrap the bottle and some may use the cups and insulate the cups. If they do the latter, suggest that they make a lid and insulate that, too.

Leave the bottle for two hours and see if it is still cold.

Encourage them to produce their own chart to record their results.

Talk about/discuss:

What conclusions can they draw from their results?

Which material was best for keeping the drink cool?

Why do they think this was?

70. Which materials keep a container of water warm the longest?

You will need: small bottle of very warm tap water per group; plastic cups; selection of materials, including bubble wrap, cotton wool, fur fabric, felt, bin liner, tinfoil, newspaper etc; timers; sensors.

Ask the children for ideas on how they could keep a small bottle of water warm.

Split the class into small groups and give a small bottle of warm water to each group.

Encourage them to write down what they would do and what apparatus they would use.

Allow time for them to experiment.

✦ Wrap the bottle in the material.

Leave for two hours.

Is it still warm?

If possible, use sensors/data logging to track changes in temperature.

Encourage them to produce their own chart to record their results.

Talk about/discuss:

Were the same materials as good at keeping the drink warm as they were at keeping the drink cool?

Why do they think this was?

Solids, liquids and how they can be separated

71. How can you separate sand from water?

You will need: filter paper or school paper towels; funnels; sand; water; plastic cups; tablespoons.

- Show the children the equipment available and ask them to suggest ways it could be used to separate sand from water.
- Mix three tablespoons of sand into a cup containing 100ml of water.
- Place a filter paper (a paper towel folded into four will also work well) into a funnel and put into the cup.
- Be careful the funnel does not tip over the cup.
- Pour the sand and water solution into it. The water will pass through the paper and the sand will remain in the filter paper.

Talk about/discuss:

- Explain to the children that this process is called filtration.
- Tell the children that filters are like sieves with very small holes that the small pieces of sand cannot go through.

72. How can you separate solids of different sizes?

You will need: soil; stones; sand; rice; marbles; sieves; margarine tub; plates.

Tell the children that sometimes people need to separate solids of different sizes, such as stones from soil, husks of wheat from flour, or the crumbs from cereal packets.

Ask how the children think this could be done.

Explain that it is possible to use sieves.

Demonstrate how to make a sieve out of a margarine tub by piercing the bottom several times.

Provide an opportunity for them to investigate how they could separate some of the above materials using different-sized sieves.

Talk about/discuss:

What did they find out?

Encourage them to write up their experiments.

73. What happens to salt when you mix it with water?

You will need: plastic cups; salt; kettle; water; teaspoons; jugs.

- Give the cups, some salt and some water to the children and tell them that they are going to find out what happens when they mix salt and water together.
- Ask them to make a prediction and encourage them to write it down.
- Measure 100ml of cold water.
- Add a teaspoon of salt and stir.
- What happens?
- Were they correct?
- Do they get the same results if they use hot water from a kettle?

Talk about/discuss:

- Explain that the salt has dissolved into the water.
- Encourage the children to write descriptions of their observations.

Safety:

- Make sure the children stand well back from the kettle. An adult should pour the hot water into cups. Remind the children that the water is very hot.

74. Is there a limit to how much salt will dissolve in water?

You will need: salt; plastic cups; jugs; water; kettle; spoons; copy of chart.

- Explain to the class that they are going to see how much salt they can dissolve in 100ml of water.
- Give each group a plastic cup and some salt.
- Ask them to measure 100ml of water into the cup.
- Ask the children to spoon the salt into the water and count how many spoons of salt they can put in the water before it stops dissolving.
- Repeat the experiment using warm water from the tap.
- Repeat the experiment again using boiling water which has been allowed to cool slightly so that it is not scalding hot.
- Does the temperature of the water make a difference?
- Record your results in a chart like the one below:

No. of spoons of Salt	Cold water	Warm water	Hot water

Talk about/discuss:

- Explain that when the salt will no longer dissolve, it is because the water is saturated.

Safety:

- Be careful with very hot water. An adult should pour the hot water into the cups. The water should never be scalding hot. Wait for it to cool slightly before pouring it.

75. Can salt be separated from water?

You will need: salt; water; saucers; spoon; plastic cups; sunny window sill; cool cupboard; saucepan; hob.

Mix salt into water in a cup until no more can be dissolved (ie making a saturated solution). It may be possible to use the solutions the children made for the previous experiment.

Pour even amounts of the salt water solution into two different saucers.

Place one saucer on a sunny window sill and the other in a cool, dark place.

Leave the saucers until all the water has evaporated from one dish.

Look at what has been left.

Discuss why there was a difference in the rate at which the water evaporated.

Talk about/discuss:

Tell the children that the salt has been separated from the water by evaporation.

Tell them that the same result would have occurred if the water had been boiled in a saucepan, and demonstrate. Ensure the children keep well back from the boiling water.

Point out that the white speckles in the saucepan are salt.

✦ Was this faster? Why?

76. Can sugar be separated from water?

You will need: sugar; warm water; saucers; spoon; plastic cups; sunny window sill; cool cupboard; filter paper or paper towels; funnels; saucepan; hob.

Mix a sugar and water solution.

Explain that the sugar is still in the water, even though they can't see it.

Ask the children to wash their hands and then let them taste the water by dipping their fingers.

Ask the children if they think the sugar can be separated from the water.

Tell the children they are going to investigate if sugar can be separated from the water in three different ways: evaporation, heating and filtration.

The children should write down their predictions for all three methods.

Split the class into three groups. One group should:
- Pour even amounts of solution into saucers and leave in a sunny or warm place.
- Pour sugar solution through filter papers.
- Work with an adult to heat the solution on the hob.

The children should list the apparatus they need to use.

Were their predictions correct?

Talk about/discuss:

- Each group should report back to the class about what they found and show the other children their results.
- Explain that some solids, like sugar, dissolve in water but cannot easily be separated out again.
- Explain that filtering can't separate the dissolved sugar because the sugar particles are too small.

Safety

- Ensure that the children all wash their hands before tasting the solution and that all the utensils and surfaces have been thoroughly cleaned. Remind the children who are working with an adult, to heat the solution on the hob, that the hob gets very hot, and that they should not touch it.

77. Can you mix oil and water?

You will need: cooking oil; plastic bottle with a screw-on lid; washing-up liquid; water.

- Pour one cup of water and a thin layer of cooking oil into a plastic bottle.
- Screw the lid onto the bottle and shake it.
- Do the oil and water mix together?
- Add a squirt of washing-up liquid.
- Put the lid on and shake the bottle again.
- What do the children notice? Are the oil and water still separated?

Talk about/discuss:

- Explain that oil and water are both liquids but will always separate into two layers, no matter how hard, or long, you shake the bottle. This is because water particles are attracted to each other but not to the particles of oil, so the water particles don't combine with the oil. They separate into different layers because different liquids have different densities, which means that some liquids are heavier than others. Oil is lighter than water, so it floats on top.
- When the washing-up liquid is added, it breaks up the surface tension of the water, which allows the water to mix with the oil.

Gases around us

78. What happens to air when it heats up and cools down?

You will need: hot water; 2-litre plastic bottle with a screw-top lid; balloon; ice; tape measure; bowl; timer.

This experiment works better when demonstrated by the teacher, with the children observing and drawing annotated diagrams to explain what is happening to the bottle and the balloon.

First, half fill the plastic bottle with hot water.

Tip the water back out and quickly put on the lid. Observe what happens to the bottle.

Now blow up a balloon and measure its circumference with a tape measure.

Place the balloon in a bowl of hot water for 5 minutes.

Measure the circumference again. What has happened?

Next, place the balloon in a bowl of ice for 5 minutes. What happens to the balloon?

Talk about/discuss:

Tell the children that the hot water heats up the air inside the bottle. As the air cools down, it takes up less room inside the bottle and so the sides of the bottle are pushed in by the force of the air pressure outside the bottle.

Similarly, the air inside the balloon expands with the heat and then contracts when chilled.

79. What happens to puddles of water when the rain stops?

You will need: puddles; rulers; chalk; timers.

- ✦ Ask the children what happens to puddles in the playground when it stops raining.
- ✦ Allocate a puddle to each group.
- ✦ Chalk around the puddles and measure them.
- ✦ Set the timer, and after 15 minutes check the puddles again, chalk around the puddles and measure them again.
- ✦ Where has the water gone?
- ✦ Encourage the children to make annotated drawings to describe what has happened.

Talk about/discuss:

- ✦ Explain that the water has evaporated. Sometimes, if the ground is warm, you can see the steam rising from the puddles.

80. How can you compare the different amounts of air in different soils?

You will need: a selection of soil types, such as clay, chalk, peat, sand, gravel-based soils; plastic cups; jugs of water (enough for each group to have one).

- Place equal quantities of at least three different soils into three different plastic cups and label them 1, 2 and 3.
- Pour water into one of the cups of soil until the water remains level with the surface of the soil.
- Measure the volume of water that sinks into the soil by subtracting the volume of water left in the measuring jug.
- Repeat for the other two cups of soil.
- Repeat the investigation to confirm the results using the same quantity of dry soils and clean cups.

Talk about/discuss:

- Explain that the cup that uses the most water has the most air in the soil.
- Tell the children that the bubbles appear because the water pushes the air out of the spaces between the particles of the soil.

Changing state

81. Will an ice cube float in salt water?

You will need: ice cubes; jug of water; salt; spoon; stirrer.

- Discuss how the children can make this a fair test.
- Encourage the children to make a prediction first and ask them to write it down.
- Place an ice cube carefully into the jug of unsalted water and observe.
- What happens?
- Ask the children to record their observations with simple diagrams.
- Add a few spoons of salt to the jug of water.
- Stir until the salt has fully dissolved.
- Place another ice cube carefully into the jug of water and observe.
- Did the ice cube float?
- Ask the children to record their observations with simple diagrams.

Talk about/discuss:

- Tell the children that ice floats on water because ice is lighter than water. This may seem weird, as ice is a solid, but when the water particles bond together to form ice, the particles are further apart than they were when they were in the water. This makes the ice less dense (lighter) than the water, which is why ice floats.

- Ice floats in salt water the same as in normal tap water, but it melts more quickly. Explain that this is because salt lowers the water's freezing point. This is why salt is put onto roads in icy weather. The salt molecules bond with the water molecules, making it more difficult for ice to form.

More ideas:

- Another experiment worth trying is observing whether sugar melts ice in the same way.

82. What makes a difference to how quickly washing dries?

You will need: string; 3 paper towels; 3 cloths; pegs or paperclips; 2 hairdryers; timer.

- Set up a washing line in the classroom and another outside.
- Wet the paper towels and cloths.
- Hang one wet paper towel and one wet cloth on the outside washing line using the pegs or paperclips.
- Hang two wet paper towels and two wet cloths on the inside washing line using the pegs or paperclips.
- Use the hairdryer on one of the wet paper towels and one of the wet cloths hanging on the washing line in the classroom.
- Use the timer to check every five minutes how dry the paper towels and cloths are.

Talk about/discuss:

- What do they notice?
- Which dries fastest inside or outside? Paper towels or cloths? Those that were left alone or those that were blown with the hairdryer?
- Discuss the possible relevant factors that may have contributed to the drying speed of the paper towels and cloths, such as temperature, wind, amount of water, the way they were pegged to the washing line etc.

83. Which materials are soluble?

You will need: transparent plastic cups; sand; sugar; salt; chalk; 100ml of water in each cup; teaspoons; jugs.

- Encourage the children to set up an experiment to test which materials (sand, sugar, salt or chalk) are soluble and which are not.
- Label the cups 'sand', 'salt', 'sugar' and 'chalk'.
- Stress the importance of making it a fair test.
- Measure 100ml of water into each cup.
- Spoon the material being tested into the appropriate cup and stir.
- What happens?

Talk about/discuss:

- Do all the materials dissolve?
- Do all make clear solutions?

✦ Explain that not all materials dissolve in water.

✦ Ask the children to list the equipment they used and to write instructions with diagrams showing how the experiment was set up.

✦ Introduce the terms 'state' and 'changes of state'.

✦ Explain that when materials dissolve, it is called a change of state. There are many other different changes of state, including melting, evaporation, condensation and freezing.

More about dissolving

84. Do sugar and salt dissolve equally in water?

You will need: salt; sugar; plastic cups; jugs; water; spoons; thermometers; copy of chart.

Explain to the class that they are going to compare how much salt and sugar they can dissolve into 100ml of water.

Give each group two plastic cups and some salt and sugar.

Label one cup 'Salt' and the other cup 'Sugar'.

Ask them to measure 100ml of water into each cup. Decide what exact temperatures of water you wish to use before the experiment begins.

✦ Measure the temperature with the thermometer to ensure that the water is the same for both salt and sugar.

Ask the children to spoon the salt and sugar evenly into the water.

Count how many spoons of salt they can put in the water before it stops dissolving and how many spoons of sugar they can put in the water before it stops dissolving.

Discuss with the children whether they think the amount of stirring will make a difference to the speed it dissolves. Explain that to ensure it is a fair test, both solutions should be stirred the same number of times at the same speed.

Record their observations in a chart.

	Cold water	Warm water	Hot water
Salt (tsp)			
Sugar (tsp)			

- ✦ Are they the same?

Talk about/discuss:

- ✦ Compare each group's results and draw conclusions.
- ✦ Do they think they can have confidence in their data?
- ✦ Was it a fair test?
- ✦ Was it a big enough sample?

Safety:

- ✦ Be careful when children are using hot water. Ensure extra adults are available for supervision.

85. How can you make a solid dissolve faster?

You will need: sugar cubes; sweeteners; salt; plastic cups; jugs; water; kettle; spoons; thermometers; timers.

- ✦ Explain to the children that they are going to plan and carry out their own investigations.
- ✦ Try not to influence their ideas.
- ✦ Ask the children to think of everyday examples of solids dissolving in water, such as sugar or sweetener in a cup of tea, salt in water for cooking or washing powder in washing machines.
- ✦ Ask them to give ideas on what would make something dissolve more quickly and list their ideas, such as stirring, amount of solid, temperature of water, volume of water etc.
- ✦ Emphasize that they are trying to make the solid dissolve quicker, rather than just adding more solid.
- ✦ Tell them they must ensure it is a fair test.
- ✦ Allow each group to decide what solid they will be investigating and to choose their own apparatus.

Safety:

- ✦ Be careful when children are using hot water. Ensure extra adults are available for supervision.

More ideas:

- ✦ Record findings in charts of their own design.
- ✦ Encourage the pupils to write sentences about what they found out and whether their predictions were correct.

86. Will steam condensed from boiling blue ink be blue?

You will need: blue ink; hob; old saucepan; oven glove; white tea plate.

- ✦ Ask the children to predict whether steam condensed from boiling blue ink will be blue or not.
- ✦ Demonstrate by boiling the ink on the hob in an old saucepan.
- ✦ Hold a white tea plate over the steam to see what colour it is.
- ✦ Ensure you use an oven glove.

Talk about/discuss:

- ✦ Explain that only the water has evaporated, leaving the ink behind.

Safety:

- ✦ When boiling liquid, keep children well back.

Enquiry in environmental and technological contexts

87. What is an effective way to clean dirty water?

You will need: muddy water; filter papers; plastic cups; cling film; A1 paper; marker pens.

Explain to the children that they are going to plan and carry out their own investigations.

Split the class into small groups. Ask how they could clean dirty water. List their ideas.

Look at the equipment available and help them plan the investigation on A1 paper. Try not to influence their ideas. Some may decide to use evaporation and some may decide to use filtration.

Allow time for them to conduct the investigation.

Talk about/discuss:

Explain that both evaporation and filtration are suitable methods, but with evaporation they will need to consider ways to collect the water.

Discuss ways of collecting water, such as covering the cups with cling film. Ask how they will stop it dripping back into the muddy water. Suggest they could have used flasks linked by a tube that dripped into another flask.

Did the filtered water look clean?

Ask the children which method they thought produced the cleanest water. What could they do to get really clean water?

88. How much salt does it take to make an object float?

You will need: salt; teaspoon; bowls; water; stones; marbles; apples; eggs; paperclips; stirrers.

- First, fill each of the bowls with 2 cups of water. Add one item in each bowl. (Either do one bowl/item at a time using fresh water each time or show all five together.)

- Add one teaspoon of salt at a time and stir.

- Count to see how many teaspoons of salt it takes to make each item float.

Talk about/discuss:

- Explain that the density of the object determines how much salt it would take for the object to float or whether they would float at all. Adding salt to the water makes the water denser.

- Tell the children an object floats when the water is denser than the object.

89. Which material makes the best boat?

You will need: variety of different cards and papers, such as acetates, cellophane, paper towels, kitchen towel, tinfoil etc; wax crayons; paint with small amount of PVA glue added; clear sticky tape; large bowl of water.

Tell the children they need to use their knowledge of different materials to select those they think would be suitable for making a boat. Talk about materials that soak up the water and suggest that these would not be suitable for a boat.

Ask the children how they could use the wax crayons or paint to improve their boat.

Remind the children about waterproofing.

Children might also need reminding that the paint will have to dry before they can test their boat, and they should be encouraged not to apply it too thickly.

Allow time for the children to investigate their own ideas.

Children should draw diagrams of their boats, with explanations of why they chose that material.

Talk about/discuss:

Encourage them to give reasons for their choice of material.

Why didn't they think kitchen towel or paper towel would be as good?

How did you use the paint or wax crayon to make your boat even better?

Display their diagrams and explanations with their boats even if the boat was not successful.

Reversible and irreversible changes

90. What happens when you heat eggs?

You will need: eggs; hob; saucepan; plate.

- Crack two eggs onto a plate and allow the children a chance to draw labelled diagrams of them.
- Heat them slowly in a saucepan (preferably not the one used to boil the ink in).
- Encourage the children to draw labelled diagrams of the eggs again just as they begin to solidify.
- Cook the eggs until they are very firm and ask the children to draw them again.
- Is the change reversible?

Talk about/discuss:

- Explain that heating some materials can cause them to change.
- Also, point out that heating is not the same as burning.

Safety:

- Remember that some children may be allergic to eggs, so do not leave them lying around in the classroom. Dispose of the cooked eggs and shells safely at the end of the experiment.

91. What happens when you burn wax?

What you need: candles; matches; trays of sand.

- Ask the children to predict what will happen to the wax candle when it is burned.
- Explain that the candle needs fuel (wax) and oxygen to keep burning.
- Provide a candle for each group.
- They should weigh the candle and note the weight.
- Place the candle in a sand tray and light it.
- When it is half of its original size, put out the flame and weigh the candle again.
- Relight the candle in the sand tray.
- When the candle goes out, collect anything that is left and weigh it.
- Discuss what happened to the wick.
- What has happened to the wax?
- The children should record their observations at each stage.

Talk about/discuss:

- Explain that when materials are burnt, new materials are formed, and in this case some of the wax has turned to gas.

- What were their findings as they weighed the wax?

- Point out that changes which occur to materials when they are burned, are not reversible.

Safety:

- Discuss safety issues with the children in relation to burning. Explain that the wax will also be very hot and and will not only burn your skin but will stick to your skin in the process, so allow the candle to cool before weighing it.

Light and dark

92. Can you see objects in the dark?

You will need: shoeboxes with lids; selection of objects young children would recognize (eg toy car, small teddy, plastic brick, crayons, pencil sharpener etc); torches.

- ✦ Have enough shoeboxes for one per group.
- ✦ Make a hole in the middle of the narrower side of the box so the children can shine a torch through, and a smaller hole in the longer side for the children to look through.
- ✦ Allow each group to choose a selection of three or four objects to be placed in the box.
- ✦ Each group should challenge another group to recognize the objects they have chosen.
- ✦ Encourage the groups to take it in turns to place an object into the box.
- ✦ Remind the children to keep the other objects hidden.
- ✦ Keep the box stationary on a table so the objects don't move, one child should look into the spy hole and predict what they think the object is before they turn on the torch.
- ✦ Next, another child shines a torch in whilst somebody from another group looks in to see what the object is.
- ✦ Allow them to have three guesses as to what the object is before it is removed from the box and they are shown what it is.
- ✦ Which group recognized the most objects correctly?

Talk about/discuss:

- ✦ Explain that when there is no light, an object is hard to see, and a torch will enable an object to be seen more clearly.

93. How can you be seen in the dark?

You will need: white shirt; reflective strips; luminous strips; torch.

Ask for three volunteers. Give one child a white shirt to wear, another some reflective strips and the third child some luminous strips.

Switch off all the lights and close the blinds and/or curtains to make the room as dark as possible.

Ask the class whom they can see.

Shine the torch in the direction of the volunteers but be careful to avoid their eyes. Whom can they see now?

Remove the sets of clothing one at a time from each volunteer. After each one, ask whom they can see now.

Ask the children to suggest ways they can make themselves seen in the dark.

Talk about/discuss:

Explain the difference between reflective and luminous materials. Also explain the importance of wearing such materials in the dark, for example when walking to school early in the morning or going home at night.

Establish that, even though an object may be shiny, it isn't a light source.

Help the children to recognize that a shiny object needs a light source for it to shine.

Light and shadows

94. How are shadows formed?

You will need: torches; selection of items such as wide-toothed combs, paperclips, card cut in various shapes, hands etc.

- Let the children explore shadow formation.
- Give them a selection of items and ask them how they can see the shadows.
- Ask the children to record what they see by drawing diagrams and writing explanations.

Talk about/discuss:

- Explain that shadows are formed when light travelling from a source is blocked.
- Encourage the children to design and make their own shadow puppet show.

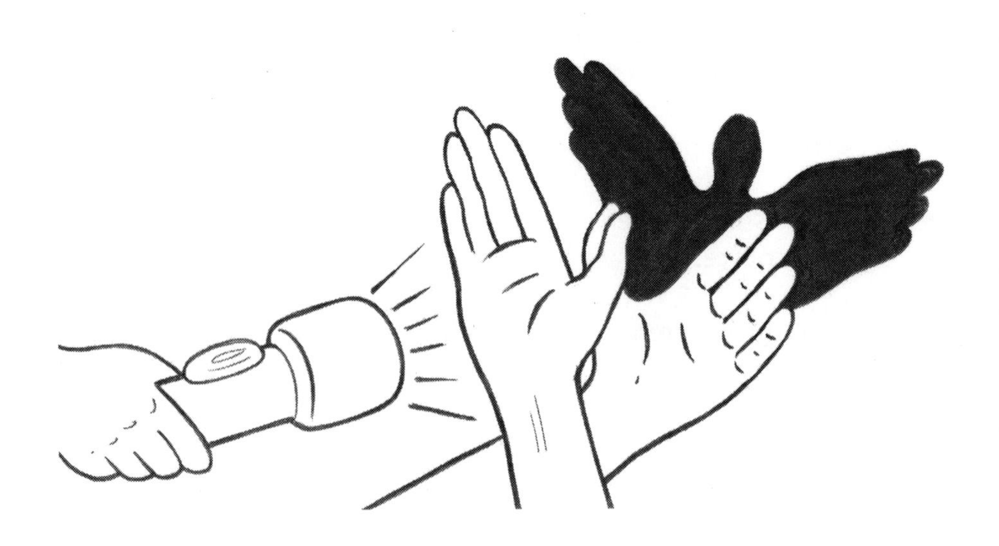

95. Do shadows change in length and position?

You will need: chalk; playground; sunny day; timer.

Split the class into pairs.

Ask one child from each pair to stand still in a space on the playground whilst their partner draws around their feet and their shadow using chalk.

Every hour during the school day, ask the children to return to their original places, using the feet marks to stand in the exact same position, and their partner should draw around their shadow again.

What has changed?

Talk about/discuss:

Explain that shadows change in length and position throughout the day.

Measure the lengths of the shadows and produce bar charts to show the length at different times of day.

Discuss their findings.

Predict where the shadow will be in three hours.

Safety

If children need to stand outside in the sun for long periods, remind them to wear sun lotion and hats. Also, they should have access to water.

96. Do shadows move in regular patterns each day?

You will need: bamboo cane; string; scissors; nails; masking tape; field; sunny week; timer.

Push the bamboo cane upright into the field in a place where it will not be touched or tampered with during the day and where it will always be in sunlight.

Mark north, south, east and west on the grass with masking tape.

Every hour, starting at 9am, ask a child from the class to mark the shadow with the string. Push the string into place by tying each end to long nails.

Label the 9am string, 12 noon string and 3pm string.

Emphasize the importance of not moving the other pieces of string.

Over the next few days, continue to check the stick to see if the shadows are in the same places at the same times.

Talk about/discuss:

Ask the children what patterns they can see.

Ask the children to explain what these observations mean.

Explain that they are observing the movement of the sun.

97. What materials throw the best shadows?

You will need: selection of items such as tracing paper, card, tinfoil, sponge, net, coloured acetate sheets, plastic bottles, greaseproof paper, paper bags, tights etc; torches; copy of chart.

✦ Give the children a selection of materials and ask them to predict which material will produce the best shadow.

✦ Complete a chart like this:

Material	Prediction	Result

✦ Test each material in turn by shining a torch on to it and observing the shadow it produces.

✦ Do their results support their predictions?

More ideas:

✦ Write an account of what they did and what they found out.

How we see things

98. Do shiny surfaces reflect light better than dull surfaces?

You will need: mirrors; polished metals; Perspex® (coloured panels); paper; gloss and matt painted surfaces; polished wood; torches.

- ✦ Ask the children to find out which surfaces they can see themselves in and which surfaces reflect a torch beam.
- ✦ Shine a torch on each surface in turn.
- ✦ Encourage the children to record their results by producing their own chart of observations.

Talk about/discuss:

- ✦ Write sentences to show generalizations about shiny surfaces, such as polished surfaces reflect light better than other surfaces, shiny surfaces can be used as mirrors whereas dull surfaces can't.

Safety

- ✦ Ensure the mirrors do not have sharp edges.

99. How can you make light bend?

You will need: mirrors; torches; cardboard; masking tape.

✦ Explain to the class that they are going to explore what happens when light hits a mirror at different angles.

✦ Discuss the angle the mirrors need to be at to make the light reflect onto a piece of cardboard. Pictures of targets can be drawn on the card if desired.

✦ Can they make a beam of light move around the whole classroom?

✦ Draw diagrams with arrows showing the direction of the light.

Talk about/discuss:

✦ Explain that light from an object can be reflected by a mirror, the reflected light enters our eyes and we see the object.

✦ If possible, examine how a periscope works.

✦ As a design-and-make task, children could make their own periscopes.

100. How can you make your shadow bigger?

You will need: overhead projector; white chalk; A3 black sugar paper; white art paper; chair; wall; Blu-tack®.

✦ Challenge the children to fill an A3 black piece of sugar paper with the silhouette of the side view of their face.

✦ Stick the black sugar paper to the wall with Blu-tack®.

✦ Each child in the group should take it in turns using the overhead projector and allocate a member in their group to move and operate the projector, draw around a silhouette, sit in the 'hot seat' to have their shadow enlarged and cut out a silhouette.

✦ Repeat until everyone in the group has had a turn at all tasks.

✦ The silhouettes can then be stuck on a larger sheet of white art paper for display within the classroom.

✦ Remind each group that they should not tell the next group how they did it, so that all groups can investigate the phenomenon for themselves.

Talk about/discuss:

✦ Ask the children to list the factors they found which affected the size and position of their shadows.

✦ How did changing one factor cause their shadow to change?

Safety:

Caution the children that parts of a projector, especially the bulb, can become very hot, and to take particular care not to touch these parts. If necessary, have an adult adjust the projector at the children's direction.

Pushes and pulls

101. Which parts of your body move?

You will need: school hall; PE kits.

- ✦ As part of a PE lesson, ask children to investigate which parts of their body move.
- ✦ Working with a partner, children can count how many joints are in their bodies.
- ✦ Children should mirror each other's movements, trying to use all the joints they identified.
- ✦ Split the class in two and have half the class watch the others mirror movements and comment on which one moved the most body parts.
- ✦ Swap and repeat with the other half of the class.

Talk about/discuss

- ✦ Ask the children to describe different ways of moving.

102. How many things in the classroom can you push or pull?

You will need: labels saying 'push', 'pull' and 'push and pull'; classroom; copy of the chart.

- Each child should work with a partner.
- Challenge them to find things in the classroom that need either a push or pull to make them move, or both, such as chair, door, drawer, piano keys etc.
- Ask the children to stick labels of 'push', 'pull' and 'push and pull' on the relevant objects.
- Copy and complete the table below.

Item	Push	Pull	Push and pull

Talk about/discuss:

- Ask the children if there are any objects that could be moved without being pushed or pulled.

Safety

- Care is needed when moving heavy objects.

103. How can you stop a moving object?

You will need: wind-up toy cars (enough for each group); selection of materials the children could use to stop movement, such as bubble wrap, cushions, fabric, bricks.

Wind up the toy car and ask the children to suggest ways they could stop it moving before it stops itself.

Let the children test their ideas.

Did their ideas work?

Would they be able to use the same methods on a real car?

Discuss what things are used to stop real vehicles in emergencies, such as if the brakes fail, you might use the gears, sand, gravel, water bags etc.

Talk about/discuss:

Explain why it would be dangerous to stop some moving objects.

Discuss road safety.

Forces and movement

104. How can you make a soft ball change direction?

You will need: soft balls; selection of bats and rackets; playground; PE kits.

- As part of a PE lesson, ask the children how they could make a soft ball change direction.
- Using their ideas, the children should hit the soft balls to each other.
- Classify their action as a push or pull.

Talk about/discuss:

- Extend the investigation to look at how they can make the soft balls move faster and slower.
- Back in the classroom, ask the children to write sentences to describe how they made the soft balls speed up, slow down and change direction.

105. Do bigger toy cars travel farther than smaller toy cars?

You will need: toy cars in a selection of sizes; metre rulers; art straws.

- Split the class into groups.
- Give each group two toy cars of different sizes.
- Ask the children to investigate which toy car goes the farthest on a flat surface.
- Tell them to push the cars with equal force and measure how far they go with the metre rulers or art straws.
- Tell the children to repeat the experiment several times to see if they get the same result each time.
- Which went farther?
- What influenced how far it went?

Talk about/discuss:

- Were their comparisons fair?

106. What elements affect the distance a toy car travels after leaving a ramp?

You will need: toy cars; ramps; lengths of track; metre rulers; tape measures.

- ✦ Ask the children what makes a difference to how far the car travels from the bottom of the ramp.
- ✦ Let the children decide how they will measure how far each car has gone.
- ✦ Encourage the children to give one variable that will influence how far the car travels, such as height of ramp, how far up the ramp the car starts, surface of ramp and amount of push, if any, etc.
- ✦ Test the cars and keep a record of how far they travel by measuring from the bottom of the ramp to where the car stops.

Talk about/discuss:

- ✦ Were their comparisons fair?
- ✦ Describe one way the comparisons might not have been fair.
- ✦ Draw bar charts to show which car went farthest.

Magnets and springs

107. What happens when magnets are put together?

You will need: selection of magnets (such as wand, bar, horseshoe, circular, ceramic etc) with both ends clearly labelled as to polarity.

- ✦ Allow time for the children to experiment with the magnets and see for themselves what happens when they are put together.
- ✦ Ask them to try the magnets from different ends.
- ✦ Encourage them to record their findings with relevant diagrams.

Talk about/discuss:

- ✦ Explain that there are forces between magnets and that the magnets attract (pull toward) and repel (push away from) each other.
- ✦ Remind children that pushes and pulls are examples of force.

108. What materials are magnetic?

You will need: magnets; a selection of materials found in and around the classroom.

- Ask the children to suggest which materials are magnetic and which are not.
- How can they find out whether they are right?
- Allow them to investigate with the magnets.
- Encourage the children to devise their own way of keeping a record of their findings.
- Classify the materials into groups such as wood, plastic, iron, copper and aluminium.
- Make generalizations about their magnetic behaviour, such as 'Wood is not magnetic', 'Iron is magnetic' etc.

More ideas:

- Ask the children to use secondary sources to find out about everyday uses of magnets, for example door catches, purse clasps, fridge magnets etc.

109. Which magnet is the strongest?

You will need: selection of different-sized magnets; paperclips.

- Using an assortment of magnets, encourage the children to find out which magnet is the strongest.
- Ask the children to design an experiment and choose which variable they will measure:
  - How many paperclips the magnet will pick up.
  - The distance the magnet will attract a paperclip.
  - How many paperclips a magnet will hold end to end.
- Allow at least 20 minutes of exploration.
- Ask for volunteers to demonstrate how they measured the strength of their magnets.

Talk about/discuss:

- Discuss their techniques with the class and ask others what they did.
- Discuss the fairness of their experiments.
- Discuss how there are many ways to do things, and explain that sometimes we learn only by trying out lots of different ideas.

110. Do magnets work through materials?

You will need: magnets; an assortment of materials, such as paper, card, wood, trays, table top etc; paperclips.

- ✦ Using an assortment of magnets, encourage the children to find out if magnets work through materials.
- ✦ Ask the children to design an experiment and choose what variable they will measure, such as through how many sheets of paper a magnet will pick up a paperclip or how thick the material can be before the magnets stop working.

Talk about/discuss:

- ✦ Discuss if their investigations were fair or not. For example, did they use the same size paperclips or the same magnet each time?
- ✦ Encourage them to record their findings and explain what their results show.

111. How far can you catapult a toy car?

You will need: elastic bands; nails; toy cars; flat pieces of wood; metre rulers; tape measures; 30cm rulers; copy of chart.

- Show the children how to make a catapult using elastic bands to propel a toy car along a flat surface, by securing an elastic band around two nails fixed to a piece of wood. For bigger toy cars, a chain of elastic bands could be looped around the legs of a sturdy chair.
- Ask them to predict what will happen if the bands are stretched by different amounts. Measure the amount of stretch to the elastic band with a centimetre ruler.
- Ask questions to help the children to decide how to test their predictions, such as:
  - What would stay the same in the experiment? (same elastic band, same toy car)
  - What would vary? (the amount of stretch)
  - What would they measure? (amount of stretch and the distance travelled)
  - How would they measure it? (from where to where, using a tape measure or ruler)
- Ask the children to record measurements in a chart similar to the one on the next page.
- Suggest the children repeat each amount of stretch several times to see if they get the same result.
- Discuss why results may vary.

Amount of stretch	Distance travelled

Talk about/discuss:

- Look for patterns in the measurements.
- Why did stretching the elastic band more make the toy car move farther?
- Explain that the more the band is stretched, the bigger the force acting on the object being propelled.

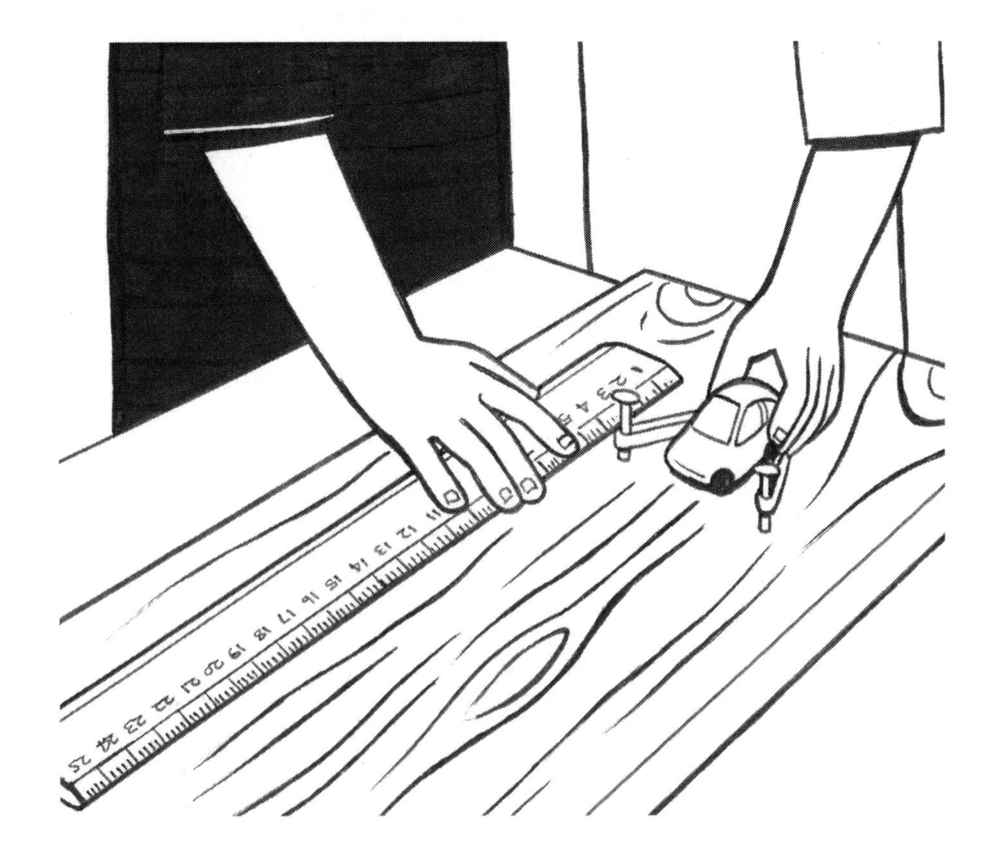

Friction

112. On which surfaces do objects slide more easily?

You will need: boards with different surfaces, eg polished wood, chipboard, vinyl, carpet, plastic, linoleum etc; bricks; newton forcemeters; metre sticks; copy of chart.

✦ The children should make a prediction about which surface they think will cause the most and the least friction and why.

✦ Demonstrate how to measure how much force is needed to start a brick moving over the different surfaces, using a Newton forcemeter.

Tell the children they should measure the height of the slope and try to keep it the same each time.

Suggest the children try the brick on the same slope several times to see if they get the same result. They could record their findings in a chart like the one below:

Surface	Test 1	Test 2	Test 3	Total

Talk about/discuss:

✦ Explain that the bigger the reading on the forcemeter, the more difficult it is to get an object moving.

Encourage the children to write sentences describing their findings, such as, 'The brick needed the least force to move it over the ... surface because ... '.

Can you think of an everyday situation where ...
1) high friction is useful? 2) low friction is useful?

113. Which shoe has the best grip?

You will need: assortment of shoes with different soles; surface boards with the same surface; metre sticks; tape measures; copy of chart.

- Split the class into small groups.

- Discuss how they would devise a test to find out which shoe has the best grip. Each group should consider:
  - Where they will place the shoe.
  - Whether they should start with the toe or the heel.
  - How far the shoe will need to move.
  - What measurements they will need to make.

- Suggest they place each shoe at the top of the slope and see how far they slide down to see which one has the best grip.

- Encourage them to list the variables that will need to be kept the same. They could use a chart such as the one below:

We change	We measure	We keep the same

Talk about/discuss:

- When would shoes with high-friction soles be useful?

- When would shoes with low-friction soles be useful?

114. Which shapes move easily through water?

You will need: litre bottles full of water; Plasticine®.

- Discuss why fish and boats can move easily through water. What shape are they?
- Show the children the litre bottles full of water and ask them how they could use these with the small pieces of Plasticine® to find out which shapes move easily through water.
- Suggest they measure the time it takes for the Plasticine® to reach the bottom of the bottle.
- Ask how they could make it a fair test.

Talk about/discuss:

- Encourage the children to devise and record their results in a chart.
- Interpret results in terms of the shape of the object and water resistance.

115. Does the size of a parachute affect how long it takes to fall?

You will need: old carrier bags; string; paperclips; stopwatches; mini whiteboards; marker pens; copy of chart.

- Using the mini whiteboards and marker pens, ask groups to generate ideas of what makes a good parachute and share their ideas with the class.

- Cut a variety of parachutes of different sizes out of old carrier bags and see if size makes a difference to the rate at which the parachutes fall.

- Remind the class they should ensure it is a fair test by attaching the same number of paperclips to the parachute each time.

- This investigation provides a good opportunity to repeat readings. Ask pupils if one reading would be reliable/sufficient, and encourage their ideas to repeat the readings.

- Record their findings in a chart similar to the one below:

Size of parachute	Time of fall in seconds			Average time to fall
	test 1	test 2	test 3	

Talk about/discuss:

- ✦ Ask if the children believe they have collected sufficient evidence to draw conclusions from.
- ✦ Were there any anomalies in their data?
- ✦ How could they improve their experiments?
- ✦ Discuss with the children how old carrier bags should be recycled and the importance of recycling to the environment.

Safety

- ✦ Remind the children not to put the plastic bags over their heads as this can cause suffocation.

116. Does the shape of a kite affect how it flies?

You will need: dowelling; newspaper; masking tape; string; bin liners; field or playground free of overhead wires.

- Working in pairs and using the materials available, the children should design and make their own kites.
- Compare the different kites the children have made on the field or playground and see which one flies best.

Talk about/discuss:

- Does the type of tail they add make any difference?
- Discuss the materials they used. Was their investigation a fair test? Why or why not?
- Explain that the shape of a kite will affect the way the wind flows around it (the aerodynamics), which determines how much lift and drag it gets.
- Conclude that shape does affect how a kite will fly and that different shapes will fly in different ways.

117. What shape of paper plane flies the farthest?

You will need: A4 paper; metre rulers.

- Challenge the children to design and make various paper planes from A4 sheets of paper.
- Which one flies the best?
- Measure the distance it travels with metre rulers.

Talk about/discuss:

- Explain that for a plane to fly, it needs a force pushing it up that opposes the weight pulling it down to the ground. If everything were equal, the plane would not move – it would be in equilibrium. If there were no resistance at all, the plane would not even take off.

  So, as the plane is flying, the weight or force, resulting from gravity pulls down on the plane, opposing the lift created by air flowing over and under the wing. An aeroplane wing changes the direction of the air. The top of the wing is curved, making the air above it move faster and making the air pressure above the wing lower than the pressure below it. This lifts the wing. The wings lift the plane, and the aeroplane flies. Thrust is created by the engines and opposes drag caused by air resistance to the aeroplane. During take-off, thrust must be more than drag and lift must be more than weight so that the aeroplane can become airborne. When landing, thrust must be less than drag and lift must be less than weight.

The Earth, Sun and Moon

118. How do we know that the Earth isn't flat?

You will need: globes; access to Google Earth; toy ship small enough to stick to a globe; Blu-tack®; photographs of the Earth, Sun and Moon taken from space.

- ✦ Look at the photographs taken from space.
- ✦ Ask the children how they know that the world is round/ spherical.
- ✦ When you open Google Earth, the image of the Earth rotates as it comes into focus. You can then rotate the Earth by pressing on the arrows to demonstrate it is spherical.
- ✦ Show the globes and give time for the children to explore the globes, looking at where different countries are in relation to each other. For example, where is New Zealand and where is the United Kingdom?
- ✦ Stick a little toy figurine to the edge of a landmass and move the ship away by hand to recreate the impression of the ship 'falling over the edge of the world'. Then explain that, because the Earth is spherical, it just looks as if the ship is 'falling over the edge of the world' and that the ship is just making its way across the curved face of the Earth.

Talk about/discuss:

- ✦ Tell the children that it is sometimes difficult to collect evidence to test scientific ideas and so sometimes evidence may be indirect.
- ✦ Explain that the Earth, Sun and Moon are all spheres.
- ✦ Encourage the children to make drawings of them.

119. Which is bigger: the Sun or the Moon?

You will need: photographs of the Sun and Moon taken from space; the children's drawings of the Sun and Moon; table-tennis balls; tennis balls; footballs; Plasticine®.

- ✦ Ask the children which is bigger – the Sun or the Moon.
- ✦ When we see the Sun and the Moon in the sky, why does the Moon sometimes seem bigger?
- ✦ Encourage the children to use their knowledge of space to make models of the Sun, Moon and Earth using a golf ball, tennis ball and a football.
- ✦ Which balls would represent the Earth, Sun and Moon respectively?
- ✦ Ask them to position the balls in the playground to give an idea of their relative distance apart.
- ✦ Encourage the children to demonstrate how the Earth rotates and orbits the Sun, and how the Moon orbits the Earth.

Talk about/discuss:

- ✦ Explain that the Moon looks about the same size as the Sun, even though it is really much smaller because it is closer, in the same way that an aeroplane on the ground looks bigger than one in the sky.

  Sometimes the Moon can seem much bigger, usually when it is close to the horizon. This is because the haze in the air distorts what we see, in much the same way that crazy mirrors in a Fairground distort the reflections we see in them.

120. Does the Sun move?

You will need: masking tape; sunny window in the classroom; clock.

- ✦ Mark out the path of the Sun travelling in an arc across the sky by putting small squares of masking tape on the window to show where the Sun is every half hour during the school day.
- ✦ Don't forget to write the time on the masking tape.
- ✦ Work out the time of day when the Sun is at its highest.

Talk about/discuss:

- ✦ The apparent position of the Sun does change over the course of the day, but this does not mean that the Sun is moving.
- ✦ Explain that the Sun seems to move because the Earth is rotating.

Safety

- ✦ Remind children that it is dangerous to look directly at the Sun.

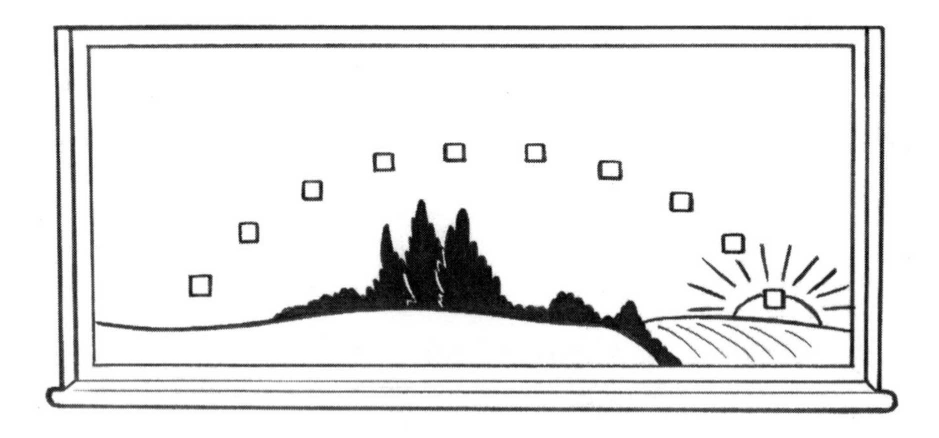

121. How long does it take the Moon to orbit the Earth?

You will need: moon-watch chart..

- Demonstrate the orbit of the Moon by asking a child to walk around a group of children representing the Earth.
- Reinforce that the same side of the Moon always faces the Earth.
- Explain that the Moon takes approximately 28 days to orbit the Earth.
- Provide each child with a moon-watch chart and ask them to complete it for homework. Make sure they include the dates.
- Tell the children they are going to keep a diary recording the phases of the Moon over the next 30 days and shade to leave the part of the Moon they see.

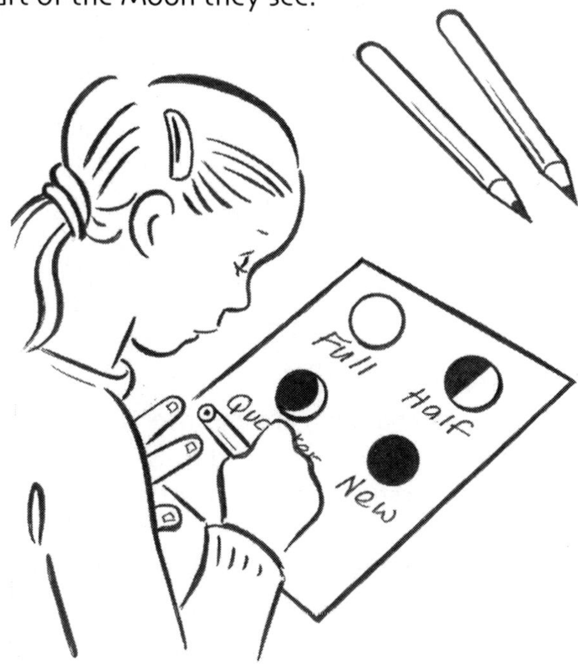

Date	Date	Date	Date	Date	Date
Date:	Date:	Date:	Date:	Date:	Date:
Date:	Date:	Date:	Date:	Date:	Date:
Date:	Date:	Date:	Date:	Date:	Date:
Date:	Date:	Date:	Date:	Date:	Date:

Talk about/discuss

- ✦ Explain that the changing appearance of the Moon over 28 days provides evidence for a 28-day cycle. Can they see how nights 29 and 30 reflect back to nights 1 and 2?
- ✦ How is the faint amount of earthshine/starshine on the dark side of the Moon (visible during the crescent stages on clear nights) evidence of its spherical shape?

Forces in action

122. Does the shape of paper affect the speed it falls?

You will need: A4 paper; stopwatch.

- In small groups, ask the children to design an experiment to test if the shape of a piece of paper will influence the speed it falls.

- Tell them they are going to test four pieces of paper, exactly the same type, size and colour. They can arrange or fold the paper in any way they want.

- Explain that they will measure the speed it falls using a stopwatch.

- Explain that they must start the piece of paper from the same point each time and that they are dropping it, not throwing it.

- Encourage them to record their findings as they go along.

Talk about/discuss:

- Allow time at the end for each group to report what they found out.

- Explain that air resistance is a force that slows moving objects.

123. Will a pendulum change direction as it swings?

You will need: 1 paper cup; scissors; string; masking tape; broom handle or dowelling; 2 chairs; sand; newspaper; large sheet of coloured paper.

Pierce a small hole in the bottom of the paper cup with the point of the scissors and seal with masking tape. The hole should be large enough for the sand to run through freely, but not so big it pours out too fast.

Pierce holes either side at the top of the paper cup with the scissors and thread the string through, so making a miniature bucket with a handle.

Balance the broom or dowelling between two chairs, with the paper-cup bucket dangling in the middle.

Place a large sheet of newspaper underneath with the coloured piece of paper on top.

Fill the paper cup to the brim with sand.

Remove the masking tape and start the pendulum swinging gently.

Let the pendulum swing until the sand has all gone and observe the patterns on the coloured paper to see if it changed direction.

Does the length of the string make a difference to the angle of swing?

Talk about/discuss:

- ✦ Explain that a longer pendulum swings more slowly than a shorter one.
- ✦ Explain that the plane in which the pendulum swings rotates slowly because of the rotation of the Earth.

124. What affects the time a spinner takes to fall?

You will need: paperclips; A4 paper spinner design; copy of chart.

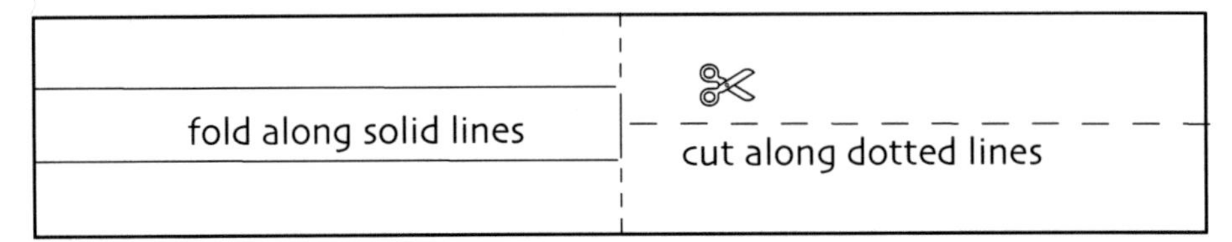

- Tell the children that they are going to predict how they could make the paper spinner fall more slowly and then test their ideas.
- Encourage them to consider the height and weight of the spinners. 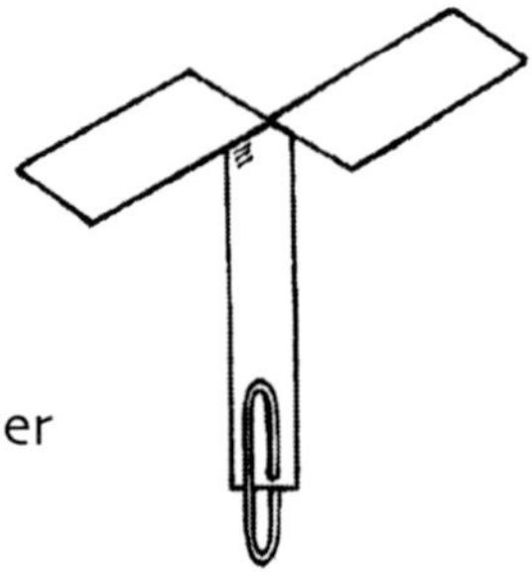
- What would happen if you made the spinner smaller?
- What would happen if you added more paperclips?
- How could you find out what the best size for a paper spinner would be?
- Encourage the children to fill in their results as they try out their ideas. Use a chart similar to the one below:

Length of wing	Time taken to reach the floor

Talk about/discuss:

- Explain that air resistance slows down the paper spinners.
- Draw line graphs using the results recorded on their charts with 'Length of Wing' on the horizontal axis and 'Time Taken to Reach the Floor' on the vertical axis.

125. How can you make a lump of Plasticine® float?

You will need: Plasticine®; large bowls of water; Newton forcemeters.

- ✦ Challenge the children to shape a small piece of Plasticine® to make it float.
- ✦ If they push down on their boats, can they feel the upthrust pushing up?
- ✦ Weigh their boats with a forcemeter in the water and out of the water. What do they notice?
- ✦ Draw annotated diagrams with arrows to show the direction of the force.

Talk about/discuss:

- ✦ Which shapes floated the best?
- ✦ Explain that the reading with the forcemeter when the boat was in the water was always zero because, when an object floats in water, its weight is balanced by the upthrust of the water. The water pushing up on their Plasticine® boats balances their boat's weight, and the force of gravity pushing down on the boat equals the force of the water pushing up.

126. How can you make cotton wool fall as slowly as possible?

You will need: cotton wool; timer.

- Split the class into pairs and give each pair a piece of cotton wool and a timer.
- Challenge them to make their cotton wool fall the slowest in the class.
- How can they do this?
- Allow time for them to investigate how the cotton wool falls.
- Tell the children that, to make it a fair test, it must be dropped each time from the same specified height of their choice.
- Encourage them to record the results of their investigation.
- When they think they have got their cotton wool to fall the slowest they think it can possibly fall, hold races where four or five pairs drop their cotton wool at the same time. The slowest from each group race against each other until you have a winner.

Talk about/discuss:

- Which pair's cotton wool dropped the slowest?
- Why do they think theirs went the slowest?
- Encourage the children to explain this in terms of air resistance.

Sound and hearing

127. From how far away can you hear someone talking?

You will need: playground.

- ✦ Split the class into pairs, child A and child B.
- ✦ In the playground, space out each pair. and ask children A to stand opposite children B.
- ✦ Encourage them to talk to each other in normal voices. Both take a step backwards and carry on talking.
- ✦ Remind the children that they must talk with normal voices and not shout.
- ✦ Can they still hear each other after five steps back?
- ✦ Repeat until they can no longer hear each other and then stop.
- ✦ Tell one child from each pair to slowly walk toward their partner, counting their steps.
- ✦ How far away were they?

Talk about/discuss:

- ✦ Compare how far away from each other the members of each pair were.
- ✦ Which pair was the farthest away?
- ✦ Emphasize that when you are far away, you cannot hear people talking.

128. How can you make sound quieter?

You will need: continuous buzzer; plastic cups; bubble wrap; cotton wool; sponges; fur fabric; cardboard.

- Ask the children what it is like to be near a loud sound.
- Let them listen to the buzzer. Ask them to come closer. Does the sound get louder or quieter?
- Ask them how they think they could make loud sounds such as the buzzer quieter.
- Suggest they wear earmuffs.
- Challenge them to make some earmuffs out of the materials available.

Talk about/discuss:

- What happens when they wear their earmuffs? Is the buzzer louder or quieter?
- Explain that the earmuffs help to stop the sound getting to their ears and that is why it is getting quieter.

Safety:

- Explain to the children that they should not listen to intensely loud music, whether it be through earphones or speakers, as it can seriously damage their hearing. Also, they should never poke anything down into their ears as this can be potentially dangerous, too.

Changing sounds

129. How can you change the pitch and loudness of a drum?

You will need: selection of different types of drums; copy of chart.

- ✦ Experiments 129, 131–132 can be conducted at the same time as a round-robin over three science sessions.
- ✦ Look closely at different types of drums, and list what is vibrating when a sound is made.
- ✦ Record the children's findings in a chart like this:

Instrument	Size	What vibrates?

- ✦ How can they change the pitch of the drum?
- ✦ Show them how pitch varies with the size of the drums and by tightening the skin with screws at the side.
- ✦ Allow time for the children to investigate how to make differently pitched sounds.
- ✦ Encourage the children to show the class what they discovered.

Talk about/discuss:

- ✦ Explain that pitch and loudness of sound are produced when an object vibrates. It depends on the size and hardness of the material that an object is made from. Explain that small drums are generally higher pitched than larger drums.

130. How can we record vibrations?

You will need: 30cm rulers; lined A4 paper; pencils; masking tape; clipboards.

Tell the children they are going to record on paper the sound of a ruler vibrating.

Clamp the end of a ruler to a table.

Using masking tape, fix the pencil securely to the end of the ruler so that the pencil is horizontal and at a right angle to the ruler.

Put the paper on a clipboard and hold it vertical against the pencil lead.

Push the ruler down and let go. Hold the clipboard to the pencil.

As the ruler vibrates, move the paper towards you. The vibrations of the ruler will be reflected on the paper by showing as zig-zag lines.

Why are some of the lines on your paper smaller? What was happening then?

Why are some of the lines larger? What was happening then?

Talk about/discuss:

Explain that sounds are made when objects or materials vibrate. The lines are longer when the sound was loudest and smaller when the sound was quietest.

Why does the sound get quieter?

✦ Label the lines to show what the volumes of the sounds were at certain points and explain why the sounds were like that.

Safety:

✦ Children should not twang the ruler too hard or it may snap.

More ideas:

✦ Demonstrate more vibrations by tapping a tuning fork and placing it in a bowl of water.

131. How can you change the pitch and loudness of a wind instrument?

You will need: selection of wind instruments; glass bottles; water; copy of chart.

- Experiments 129, 131 and 132 can be conducted at the same time as a round-robin over three science sessions.
- Look closely at different types of wind instruments and list what is vibrating when a sound is made.
- Record the children's findings in a chart like this:

Type of wind instrument	What vibrates?

- How can they change the pitch of the wind instrument?
- Demonstrate how the note from a wind instrument can be changed by changing the length of the air column.
- Show them how to make a sound by blowing across the top of a bottle and ask them to suggest what is vibrating.
- What happens if you add water to the bottle?
- Allow time for the children to investigate how to make high-pitched sounds, low-pitched sounds, loud sounds, quiet sounds etc with the wind instruments and the glass bottles.
- Encourage the children to show the class what they discovered.

Talk about/discuss:

- Demonstrate to the children the changes in sound using a selection of bottles containing varying degrees of water. Explain that when you add more water to the bottles, the pitch increases.

- Suggest that they sketch annotated drawings of how the sound changes in the bottles when they add more water.

Safety

- LEA/school guidelines on glass must be observed and extra adult supervision may be required.

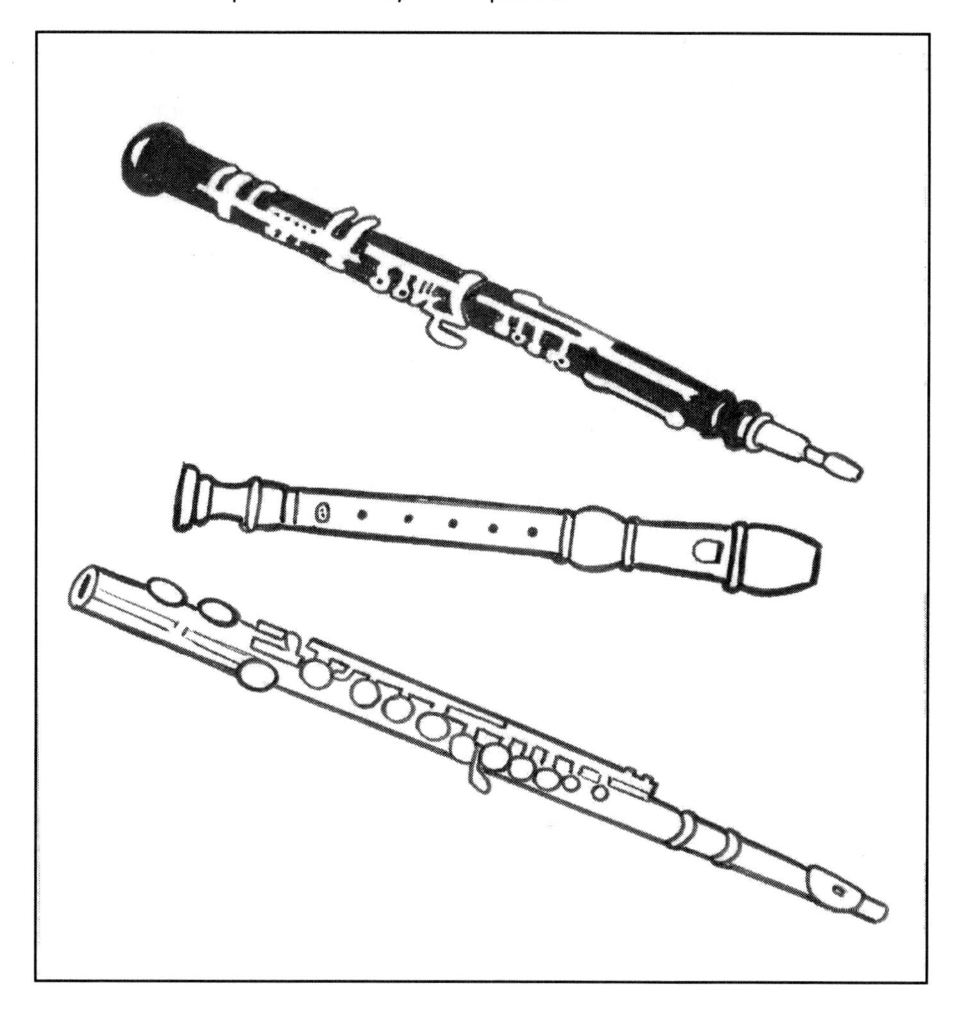

132. How can you change the pitch and loudness of a stringed instrument?

You will need: selection of stringed instruments; copy of chart similar to page 160.

Experiments 129, 131–32 can be conducted at the same time as a round-robin over three science sessions.

Look closely at different types of stringed instruments and list what vibrates when a sound is made.

Record the children's findings in a chart showing the type of stringed instrument alongside what part vibrates.

How can they change the pitch of the stringed instrument?

Show them how the note from a string can be changed by changing the length of the string, by tightening a string or by having different thickness of strings. Point out examples with the instruments available.

Allow time for the children to investigate making high-pitched sounds, low-pitched sounds, loud sounds, quiet sounds.

Encourage the children to show the class what they have discovered.

Safety:

Be careful when stretching the strings because, if over-stretched, they can break and ping onto you.

More ideas:

- ✦ If time permits, the children could design and make their own stringed instruments using elastic bands and shoeboxes.
- ✦ Can they make their own strings play different pitches?

Using electricity

133. How do you make a buzzer buzz?

You will need: batteries; battery holders; buzzers; crocodile clips; wires.

- ✦ This investigation could be carried out at the same time as the next experiment, Activity 134, except that here children make a buzzer buzz instead of a bulb light up.
- ✦ Check that you have batteries that work, and provide battery holders as this will make it easier for the children to construct their circuits.
- ✦ Wires with crocodile clips attached are less fiddly.
- ✦ Give the children the equipment and challenge them to make the buzzer buzz.
- ✦ Can they work out how to make a circuit?
- ✦ Give the children 15 minutes and then demonstrate how to make a circuit.
- ✦ Allow time for the children to construct their own circuits and show the class.

Talk about/discuss:

- ✦ Explain that electrical devices will not work if there is a break in the circuit.
- ✦ Show the children how to draw simple circuit diagrams to record their work.

134. Can you light up the bulb?

You will need: batteries; battery holders; bulbs; crocodile clips; wires

- This investigation could be carried out at the same time as the previous experiment, Activity 133, except that here children make a bulb light up instead of a buzzer buzz.
- Again allow the children time to see if they can work it out themselves.
- Demonstrate how they can use the same circuit they used for the buzzer, but replace the buzzer with the bulb.
- Allow time for the children to construct their own circuits and show the class.

Talk about/discuss:

- Again, encourage the children to draw simple circuit diagrams to record their work.
- Give examples of circuits that do not work and ask them to explain why they don't work.

135. How can you make a doorbell?

You will need: tinfoil; card; paper; paperclips; wires; batteries; buzzers; battery holders.

Tell the children that they are going to use their knowledge of circuits to design a doorbell.

Demonstrate a simple switch by folding a rectangle of card in half.

Stick small squares of tinfoil on both sides of the inside of the card.

Connect a wire to the tinfoil and connect the battery and buzzer to make a circuit.

When the two sides of tinfoil come into contact with each other, the switch should work so the buzzer buzzes.

Allow the children to experiment with their own switches and incorporate other switches they have made.

Talk about/discuss:

The children should record their designs using circuit diagrams.

Ask why they used the materials they have chosen.

136. Can you make a toy lighthouse?

You will need: bulbs; wires; batteries; battery holders; switches; junk modelling, such as tubes, plastic bottles, cups, egg boxes, yogurt pots etc; masking tape.

- Read one of the Lighthouse Keeper stories by Ronda and David Armitage.
- Ask the children to design a working toy lighthouse with the equipment available.
- Give children time to make and test their designs.
- Demonstrate the lighthouses to the class.

Talk about/discuss:

- Ask the children to explain the importance of the complete circuit in their lighthouse.

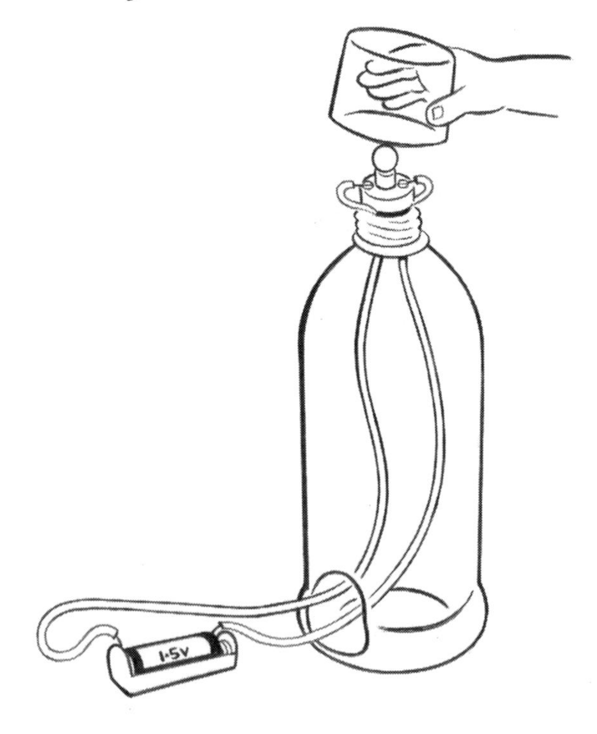

Circuits and conductors

137. How can you make a bulb brighter?

You will need: wire; bulbs; crocodile clips; batteries; battery holders; assortment of materials.

- ✦ Split the class into small groups and help to set up a circuit that comprises one bulb, wires and two batteries.
- ✦ Provide the children with extra wires, batteries and bulbs.
- ✦ Ask the children to explore ways of changing the brightness of the bulb and to consider the reasons for that change.
- ✦ How did you make the bulb brighter?
- ✦ How can you make it brighter without changing the battery?
- ✦ Why do you think the bulb got brighter?

Talk about/discuss:

- ✦ Children should record their changes using circuit diagrams.
- ✦ Use ICT simulation software to test predictions on how the addition of components in a circuit could make a bulb brighter.

138. What materials conduct electricity?

You will need: materials to check such as a pencil, wooden ruler, fork, paperclip etc; wire; bulbs; crocodile clips; batteries; battery holders; assortment of materials.

- ✦ Ask the children to predict for each type of material whether they think it will conduct electricity or not and to write down their predictions in a tick chart such as the one below:

Material	Conducts electricity?	
	yes	no

- ✦ Working in small groups, the children should make a circuit, leaving a gap for where they will insert the material they are checking.
- ✦ Some groups may need more guidance on how to do this.
- ✦ Insert each material to see if the bulb lights up.

Talk about/discuss:

- ✦ Explain that if the bulb lights up, it means the material does conduct electricity. If it does not light up, it means the material does not conduct electricity.

Changing circuits

139. Does the thickness of a wire affect the brightness of a bulb?

You will need: different thicknesses of insulated wire; bulbs; crocodile clips; batteries; battery holders; rulers.

- Present the children with the equipment available.
- Ask them to design a fair test to find out if the thickness of a wire affects the brightness of a bulb.
- Ensure they write down their predictions and list their equipment before they begin the investigation.
- How are they making sure it is a fair test? Remind them to change only one variable at a time, which is the thickness of the wire.
- Ensure the wires are the same length by measuring them with centimetre rulers.

More ideas:

- Ask children to make annotated drawings of their circuits.
- Were their predictions correct?

Enquiry in environmental and technological contexts

140. How do you make an electrical switch?

You will need: batteries; battery holders; buzzers; card; coins; crocodile clips; metal discs; paperclips; paper fasteners; sticky tape; tinfoil; wires.

- Tell the children that they are going to design and make their own switch.
- Show them the equipment available.
- Ask them to draw diagrams of their designs using circuit symbols before they make them.
- Allow time for them to build their switches.
- Demonstrate how their switch works to the class.

Talk about/discuss:

- Ask the children why they have used those materials.
- How could they have made it better?

141. How could you make a warning device?

You will need: batteries; battery holders; bumper pack of sweets; buzzers; card; crocodile clips; paperclips; sticky tape; tinfoil; wires.

- Challenge the children to design and make a warning device to tell people if a precious object has been stolen, as in a museum.
- Explain that they are going to use packets of sweets. They can work in pairs and challenge another pair to remove the packet of sweets from the card without setting the buzzer off.
- Set some ground rules in that they are only allowed to touch the packet of sweets and they are only allowed to use their hands. Limit the activity to three tries per child.
- Demonstrate one idea by showing the children how to fold the card in four like a zigzag fan, so it will spring up again when the packet of sweets is removed.
- Stick the base of the card to a table with Sellotape.
- Now stick foil to the top side of the zigzag fan.
- Connect a wire to the tinfoil and connect the battery and buzzer to make a circuit.
- Place the packet of sweets on top of the card to weigh it down.
- Hold a paperclip in a crocodile clip at the other end of the circuit. This needs to be stuck to the table so that the paperclip overhangs the tinfoil on the card.
- When the packet of sweets is taken off, the card should spring up, so the foil comes into contact with the paperclip.
- Allow time for children to experiment with different ideas.

Talk about/discuss:

- Explain that when the packets of sweets are removed from the folded card, the foil and paperclip will touch, completing the circuit and setting the buzzer off.
- Ensure that there are enough packets of sweets for every child. Let them have them to eat after the session.

Safety

- If you allow the children to have the sweets, ensure you have the necessary permissions from parents and check for any allergies.

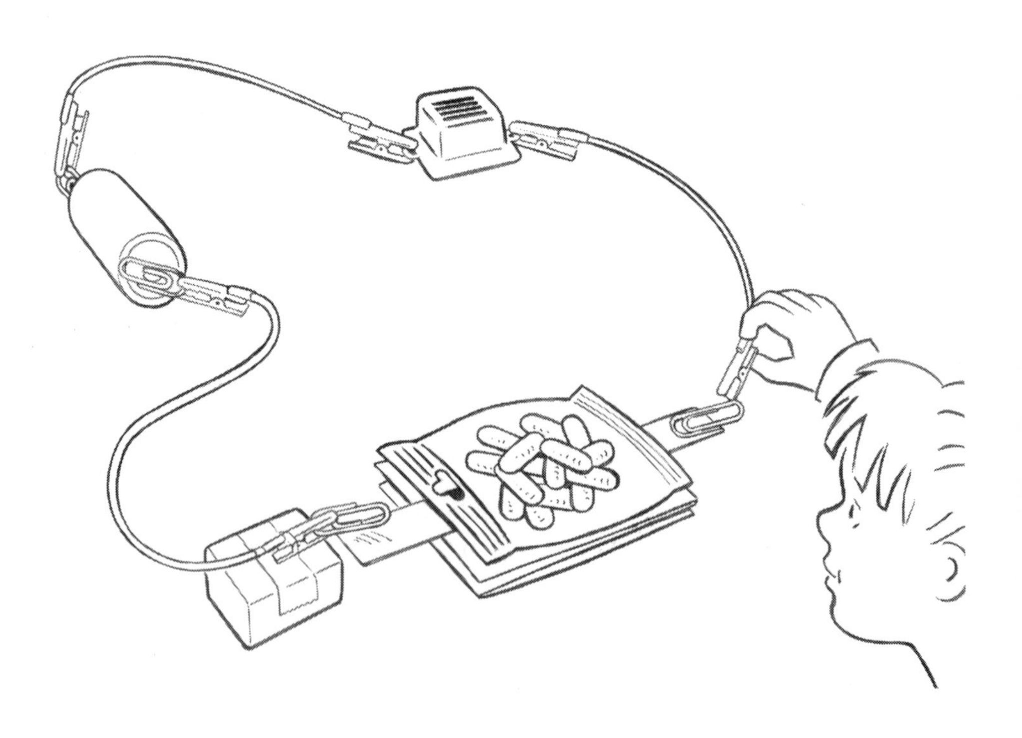

Index

Lightning Source UK Ltd.
Milton Keynes UK
09 December 2009

147274UK00001B/5/P

9 781905 780358